97 Things Every UX Practitioner Should Know

Collective Wisdom from the Experts

Dan Berlin

Beijing · Boston · Farnham · Sebastopol · Tokyo

97 Things Every UX Practitioner Should Know

by Dan Berlin

Published by O'Reilly Media, Inc., 1005 Gravenstein Highway North, Sebastopol, CA 95472.

O'Reilly books may be purchased for educational, business, or sales promotional use. Online editions are also available for most titles (*http://oreilly.com*). For more information, contact our corporate/institutional sales department: 800-998-9938 or *corporate@oreilly.com*.

Acquisitions Editor: Amanda Quinn

Development Editor Corbin Collins

Production Editor: Daniel Elfanbaum

Copyeditor: Arthur Johnson

Proofreader: Piper Editorial Consulting, LLC

Indexer: Sam Arnold-Boyd

Interior Designer: David Futato

Cover Designer: Karen Montgomery

Illustrator: Kate Dullea

June 2021: First Edition

Revision History for the First Edition

2021-05-11: First Release

See *http://oreilly.com/catalog/errata.csp?isbn=9781492085171* for release details.

978-1-492-08517-1

[LSI]

In memory of Tom "Mr. T-Test" Tullis.
Thank you for passing your love of quantitative methods to so many of us.
We lost you too soon.

Table of Contents

Part II. Strategy

Part III.   Design

Part IV. Content

Part V. Research

Preface

The user experience (UX) community is full of people who want to help others lead better lives. Whether we are making a website easier to use or creating designs that guide people to make healthier decisions, we just want to help others. This also applies to our colleagues across the industry: we want everyone who has chosen this occupation to thrive. That's why you'll often see more hugs than handshakes at UX conferences—everyone in the field is just so nice.

That's the intended spirit of this book—it's by the UX community and for the UX community, and it intends to give back to the UX community. To ensure that contributors extended beyond my personal circle, we had a submission process in which anyone could submit a chapter. This helped to solidify my theory about UXers being so nice. Every stranger I met during the process was generous and eager to give back to the community. To round out the community giving, the authors have decided on a number of UX nonprofit organizations to which any book royalties will go, starting with Creative Reaction Lab in St. Louis and the Center for Civic Design. This book wasn't written to make a buck or to give people exposure. We wrote this book to help the UX community do great work.

Many people involved with UX came across the field by chance. My UX "aha" moment came during my customer support days, when I sat as a participant in a usability study. With my psych undergrad focus on visual space perception and my love of computer geekery, this seemed like the job for me! I was even more pleased to learn you don't need to be a designer to work in UX, since I thought I didn't have a creative bone in my body. This phrase would make Adam Connor bristle in our conversations, and he eventually convinced me that *everyone* is creative—it's just a matter of how creativity is expressed by or elicited from an individual. Researchers are creative problem solvers; strategists think creatively to give design projects meaningful direction; and, as I've learned in many design studios, anyone who can draw or imagine a rectangle can design.

This book is intended for everyone in the UX community, whether advanced practitioner or beginner: designers, researchers, UX strategists, developers, content strategists, managers, and executives. The goal was to make each chapter as practical as possible and to cover topics that even advanced folks may not have previously considered. The chapters are organized into the different areas of interest in UX:

- *Career:* Tips to advance your hireability
- *Strategy:* Dealing with the fuzzy frontend of the design process
- *Design:* Creating experiences for customers, employees, and other users
- *Content:* Writing words that matter and that aren't an afterthought
- *Research:* Gathering data from study participants to uncover actionable insights

We live in a service economy. Businesses must consider the priorities of their customers, members, patients, employees, and of all other users in their ecosystem. UX strategy, research, and design are the pathways to aligning experiences with people's expectations for how things should work. People all across the business world should be aware of how the UX sausage gets made. The authors of this book hope it provides information for everyone who wants to create great experiences for other people.

Permissions

In the spirit of the first 97 Things books, each contribution in this volume follows a nonrestrictive, open source model. Each contribution is licensed under a Creative Commons Attribution 4.0 license (*https://oreil.ly/zPsKK*).

O'Reilly Online Learning

 For more than 40 years, *O'Reilly Media* has provided technology and business training, knowledge, and insight to help companies succeed.

Our unique network of experts and innovators share their knowledge and expertise through books, articles, and our online learning platform. O'Reilly's online learning platform gives you on-demand access to live training courses, in-depth learning paths, interactive coding environments, and a vast collection of text and video from O'Reilly and 200+ other publishers. For more information, visit *http://oreilly.com*.

How to Contact Us

Please address comments and questions concerning this book to the publisher:

> O'Reilly Media, Inc.
> 1005 Gravenstein Highway North
> Sebastopol, CA 95472
> 800-998-9938 (in the United States or Canada)
> 707-829-0515 (international or local)
> 707-829-0104 (fax)

We have a web page for this book, where we list errata, examples, and any additional information. You can access this page at *https://oreil.ly/97-things-ux*.

Email *bookquestions@oreilly.com* to comment or ask technical questions about this book.

Visit *http://oreilly.com* for news and information about our books and courses.

Find us on Facebook: *http://facebook.com/oreilly*.

Follow us on Twitter at *http://twitter.com/oreillymedia*, and also check out *http://twitter.com/97_Things*.

Watch us on YouTube: *http://youtube.com/oreillymedia*.

Acknowledgments

Without the support of my family, I wouldn't be where I am today. Thank you to my parents, Joel and Rita; my brothers, Eric and Adam; and the rest of the Berlin crew: Doriz, Janinne, Matt, Josh, Alex, Lea, Ruby, Mia, Sasha, Maggie, and Shadow. We may be spread across three states, but we're very lucky in how close we remain. Thanks for everything through the years.

Thank you to the reviewers who read the book and provided feedback during editing to make it stronger: Chauncey Wilson, Kyle Soucy, Bob Thomas, and Nedret Sahin. UX friends are the best friends, and I really appreciate all the effort you folks put in to help with this and other community events. Chauncey, since you ran the study that alerted me to UX as a career, your help in reviewing this book brings everything full circle and gives me much joy.

A huge thank you goes out to Tom Greever, who started this initiative and who entrusted me to edit the book. Additionally, thank you to everyone who had a submission in the initial book draft: Juan Antonio Avalos, Mudassir Azeemi, Vidhika Bansal, Rubem Barbosa-Hughes, Leon Barnard, Lea Botwinick, Shelby Bower, Luke Chen, Nick Comito, Susan Culkin, Jon Fukuda, Julie Jensen, Scott Jenson, Frank Konrad, Joseph McCarroll, Rakesh Patwari, Joe Payne, Stephanie Pi, Holly Reynolds, Calvin Robertson, Razaur Rahman Shahed, Wade Shearer, Chris Stegner, Steve Turbek, Tiffany Vurek, Caleb Williams, and Pia Zaragoza.

Thank you to my Mad*Pow family. It's been so much fun creating great experiences with you every day. You are inspiring people, and y'all have helped make me a better practitioner and person over the past 10 years.

Fast and firm to the Buffett crew: Snarf, Zoot, Edith Bunker, Mayor McCheese, Dr. Bunsen Honeydew, Pink, Henrietta, Nermal, Bozo The Clown, Cameron, Hamburglar, Zippy the Pinhead, and of course, Bomber (and the rest of the Penyacks!). You folks keep me sane and are the best friends I could ask for.

Thank you to everyone I've met and worked alongside at UXPA Boston. We're so lucky to have such a dedicated bunch of UXers in our area, and it's been so amazing and educational helping the organization through the years. Thank you for giving me the opportunity to serve the UX community and feed them delicious lemon bars.

Finally, thank you very much to the 96 other authors who contributed to the book! Without you, there would be no book and no opportunity to pass this knowledge on to the next generation of UX practitioners. Working with y'all has been a highlight of my career; everyone's positive attitude and willingness to offer their thought leadership has been nothing short of inspirational.

PART I

Career

Boost Your Emotional Intelligence to Move from Good to Great UX

Priyama Barua

The industry shift toward human-centered design and a democratized design process has encouraged UXers to optimize results by directly involving the people for whom they design. But humans are complex, and certain personal interaction skills don't come naturally to everyone. The greatest UXers I've met have one thing in common: emotional intelligence (EI).

Daniel Goleman, a recognized expert in EI, identified four competencies of EI: self-awareness, self-management, social awareness, and relationship management. I've noticed that great UXers possess a balance of the four. Let's take a closer look at how they express these competencies.

- *Self-awareness:* They're able to accurately identify their own emotional strengths and weaknesses. For example, they recognize if they're reacting emotionally to a design critique.

- *Self-management:* Their rational brain dominates their emotional brain. They're comfortable with the uncomfortable, and they keep the team moving as the messy, sometimes frustrating design process unfolds.

- *Social awareness:* They're empathetic and have strong organizational awareness, creating viable and feasible solutions with continuous inputs from business and technical partners who don't speak the language of UX.

- *Relationship management:* They're able to influence people positively and don't shy away from necessary conflict. They advocate for design to have a "seat at the table."

As I craft a path in UX, I too find myself navigating diverse priorities, opinions, and personalities, and ensuring a high degree of EI in my response.

Ethnographic research, co-creation workshops, cross-team design reviews, multidisciplinary project teams, socializing, and defending design decisions with stakeholders—all require me to reflect on how I handle situations.

Additionally, EI must extend to the people we're designing for—this means avoiding bias and truly empathizing with users' needs so products connect with them and sometimes delight them. UXers must also help stakeholders and teammates recognize users' emotions in a variety of contexts and respond with fitting design solutions.

None of this comes naturally to me; I still struggle with EI. But by learning about and practicing these competencies, I was able to strive toward UX greatness. If you'd like to do the same, here are my recommendations:

1. *Reflect on your strengths in the four EI competencies:* If you're naturally self-reflective, you can do this yourself. If not, ask your manager, coworkers, or a mentor for feedback.

2. *Identify the competencies you can hone to truly balance your EI:* Think back to situations you could have handled better. Then enlist someone you trust to help you work on these competencies.

3. *Recognize that inappropriate emotional responses are a result of an evolutionary failure:* The amygdala hijacks the rational part of our brain and kicks in a flight-or-fight response to tense situations, mistaking them for a predatory threat on our lives. The best way to overcome this is to self-talk to reframe situations, but this is easier said than done. In the words of Michael Cornwall, the author of *Go Suck a Lemon* (CreateSpace), "It will take the force of will to do that."

4. *Apply your UX skills to improve your EI:* Start tracking the outcomes you want to influence. Are your design metrics improving because of your effort? Are you called on for more leadership opportunities? Do you feel better and more in control? Last, are you able to help others grow in their EI? I've found the goal-setting framework OKR (Objectives and Key Results) helps track against outcomes like the ones mentioned here.

If you're ready to put this in practice, think about the great UXers you know. Can you recall times when you saw them exhibit exceptional EI? Emulating their qualities and being truly self-reflective about your own can go a long way in creating great UX.

Your Worst Job May Be Your Best Learning Experience

Taylor Kostal-Bergmann

It may be surprising that our industry has but a few truly mature design organizations, and as a result many UX practitioners get frustrated within the first six months of beginning a job and start looking for a new one. Often, practitioners start looking once they realize their design organization has limited resources and is working under unreasonable expectations with people who don't understand UX or its processes. They get burned out, realize that they won't be doing the inspiring portfolio work they've dreamed of, and begin to wonder if another workplace would be a better fit. Those who consider leaving fear they will end up confronting the same issues in a new company. Those who consider staying recognize they need to fight for a seat at the table, and frankly, many people just aren't interested in doing that.

I'm here to tell you that if you choose to stay, you *can* eventually change things, developing a level of skill and tenacity most designers take years to cultivate. To transform your workplace, however, you must focus on small cultural shifts rather than immediate sweeping changes. So where do you begin?

Start Small and Focus on Building Trust

I've worked in a variety of roles, and one thing I've consistently found in immature organizations is there are always some folks resistant to UX. This resistance often arises from fear or distrust, but as UX practitioners it is our responsibility to understand this and work through it. Here are some effective strategies to counteract resistant stakeholders:

- *Get to know people outside of your UX team:* Learn about their roles, performance objectives, processes, team power structure, experience with UX, hobbies, and so on.
- *Focus on individuals who seem interested in UX first:* The more advocates you have, the more help you get building confidence in resistant stakeholders.
- *Identify footholds in business processes to inject UX:* Focus on small, digestible doses that drive value without rocking the boat too much.
- *Work lean and be impeccable with your word:* This is how you build trust.
- *Serve their palate before you serve your own:* Demonstrate your value by focusing on clearly communicating user insights, the strategy they drive, and the outcomes they produce.

Diversify Your Skill Set

I was laid off from my first job after college and learned the hard way that talent alone does not ensure job security. Survival comes down to what you can bring to the table when the chips are down and your company has fewer resources to work with. You need to keep growing your skill set throughout your career. By doing this, I've survived layoffs and have been able to lean into needed skills when resources were in short supply. Some great skills you can learn are:

- Workshop facilitation (especially remote)
- User research methods
- Designing great presentations
- UX writing and content strategy
- UX strategy
- Service design

Get Organized and Say No

Many immature companies have overburdened or disorganized design teams. By focusing on workload management and boundary setting, many of these organizations are able to build trust and create a proactive relationship with product teams, helping UX thrive. Set yourself up for success by safeguarding your time and sanity with project intake requirements and kanban tools. Additionally, get comfortable saying no and educating on why.

Don't discount an immature organization's value to your career. It's fair to want to leave toxic organizations, but know the difference between toxic and disorganized. If you stay, you may find that it is extremely rewarding to help an organization transform and mature—you'll be a stronger practitioner for having the tenacity to put up a fight.

You're Never Done Learning

Andrew Wirtanen

While in high school, I loved working on websites and wanted to be a web designer. That changed a bit in college when one of my first classes was an introduction to human-computer interaction. Then graduate school allowed me to dive deeper into topics like user research and interaction design. The master's program was incredibly educational, so what else was there to know? It turns out I had only just begun a career of learning. For example, just after graduation came the release of the iPhone, and everyone had to quickly learn how to research and design apps for mobile phones.

If you don't love learning and don't have a routine to stay up to date, you risk not hearing about new ideas, methods, processes, and technology. As a result, the quality of your work may suffer and your career development may stagnate. If you need to ignite your love of learning, try exploring different topics until you find some that interest you. Don't be afraid to go down one path, only to find it boring—don't be afraid to give up on an unfulfilling topic. Once you have your interest areas, though, it's time to set up a routine.

Make Time

The most important part of your routine is to understand how much time you might have available. You may not have much time at all, and that's okay.

Some questions to ask yourself:

- Could you listen to a podcast or an audiobook while doing chores?
- Could you set up a recurring meeting with peers, colleagues, or a mentor?
- Are there conferences, book clubs, or meetups you could attend?
- Can you take microbreaks during the day to read articles?

Be Selective

We must be selective in what we learn. One natural approach to being selective is to pay attention to what's interesting or relevant to you. However, be careful! You may unintentionally ignore something that could develop into a new interest area.

Another approach, described by Vasilis van Gemert in a 2012 article, is to use people and time as filters.[1] You probably know of some people who are always sharing useful links. You could follow those people and save some of their links to check out on your own time. To filter potential noise, consider not consuming content immediately after publication—you'll find that some articles or podcasts may be outdated in a few days because they are reactions to developing news.

Refine Your Routine

One of the best ways to automate how you receive content is through subscriptions via news aggregator and podcast apps. News aggregator apps like Feedly and Pocket help you find content on the web. Instead of you checking a bunch of different websites, the news comes to you! And with a podcast app like Overcast (iOS), Pocket Casts (Android), or the default one on your phone, you can subscribe to audio episodes. I recommend increasing the playback speed and looking for a feature to skip or eliminate silences.

Remember to regularly manage your subscriptions to make sure they're still relevant. For example, you may have subscribed to a podcast related to a project. But once that project is over, the podcast may no longer be the best use of your time.

Share

When you're regularly learning and trying new things, it's important to share with your coworkers or UX community, because that will lead to reciprocal sharing, inspire new thinking, and subsequently help advance the field of UX. Whether you're working on your first project or you've been in UX for 30+ years, remember that we're all continuously learning.

---

1 Vasilis van Gemert, "How to Stay Up to Date," Smashing Magazine, August 9, 2012. https://oreil.ly/0KGdY

So You Want to Be a UX Consultant

Eva Kaniasty

Who among us hasn't dreamed of the glamorous, lucrative enterprise that is consulting? No more constraints or boredom! No more bosses or annoying coworkers!

Not so fast. For every benefit there is a drawback. Many have set out to make a living as a UX consultant, only to find themselves simply unemployed. Let me tell you everything I've learned in the last 10 years, so that you can learn from my mistakes.

- *Don't be a newbie:* Working as a consultant means learning new things every day. But those things don't include the basics of your job. Trying to do consulting right out of school is like giving your first piano recital at Carnegie Hall. Unless you already have the chops to do your job better and faster than 80% of your peers, you're not ready.

- *Be efficient:* Clients are breathing down your neck. Work is feast or famine. Meanwhile, taxes need paying, estimates need running, and your cat just threw up. Do you finish that proposal or do you clean up the vomit? Trick question—you do it all, in half the time. You probably never noticed the administrative stuff someone else handled at your day job. So learn to enjoy (and appreciate yourself for) at least some of the busywork that falls into your lap.

- *Don't be a genius:* You do need confidence to inspire trust, and a good consultant has to be smart enough to tackle wicked problems. But leave your ego at the door. A great consultant recognizes everyone's good ideas, doesn't hog credit, and is more than happy to share their knowledge. Think of yourself as a supporting actor, not the star of the show, and your clients will love you.

- *Be assertive:* Yes, you want to keep your clients happy. At the same time, as an objective outsider, you need to have the guts to speak truth to power. If you lack spine and let problems fester, you'll be working too hard for suboptimal results. Then there are the moments when you have to talk about money...loudly, clearly, and often. Get over your embarrassment or fail to get paid.

- *Don't be a loner:* Being your own boss is awesome, but it doesn't mean that you answer to no one. You are still working for and with others. A big part of finding and keeping work depends on how skilled you are at relationship building and adapting to your clients rather than forcing them to adapt to you. And that's just for your day job: add managing your professional social life to your ever-growing lists of tasks, because watercooler chat is no longer an option.

- *Be a finisher:* A project that doesn't end is a project that fails. It doesn't matter how great your personality is and how good you are at scheduling meetings—what you get done is ultimately what counts. Your clients won't tell you so, but often they are asking you simply to get things done —things they've struggled to get done internally. So if procrastination is one of your demons, banish it now, before your livelihood depends on it.

Now for the good part: consulting is great—the freedom to design your own life, the autonomy to do the work you like best, the endless novelty of new people and experiences. But if you decide to take the plunge, do it with your eyes open. And don't be disappointed if there are times in your life when you're better off keeping (and liking!) your day job.

Master the Art of Storytelling

Reena Ganga

Human beings have a deep-rooted desire to connect through stories. More than 36,000 years ago, our hunter-gatherer ancestors were recounting stories about volcanoes and wild beasts. That primitive urge to tell tales around a campfire is how we make sense of the world and foster cooperation. In today's world of tech, effective storytelling is key to driving home the value of your product, because in a world of plenty, what compels customers to buy a new device or app is the belief that this shiny new thing will fill a void they just realized they have.

As user advocates, designers and researchers have critical insight into the way customers think and feel, coupled with an intimate understanding of the product. That gives them unique power to translate the raw features and functionality into an effective story for clients and customers.

Whether you're presenting at a conference or simply on a conference call, here are my six steps for spinning a compelling yarn:

1. *Understand the audience:* First and foremost, understand what's driving your audience. A group of CEOs cares about vastly different things than an audience of developers, and only when you know *why* you're giving this presentation can you start formulating the *what.*

2. *Identify the big picture:* Zoom out. In my former career as a TV news reporter, I learned that the best stories stay with us because they strike a nerve or touch something deep inside. Harnessing emotions such as love, pain, joy, or fear can be a powerful way to reach the audience, so ask yourself if your product story is actually part of a larger issue or trend that might resonate.

3. *Capture attention:* Develop a catchy hook. A presenter has only 30–60 seconds to engage audience members before their attention drifts and they begin scrolling on their phones, so open by giving your audience a

reason to care about what you're going to say. This could take the form of a rhetorical question, a surprising statement, or even a prop. Bill Gates once released a jar of mosquitos into a TED audience to emphasize the importance of fighting malaria. It brought a faraway problem into sharp focus.

4. *Take people on a journey*: Share a use case. Every great story has a hero, and the product user is yours. Start with an obstacle, a decision point, or an ethical dilemma and take the audience on the journey of how your product solves the user's problem and gives the hero a happily ever after. However, remember to limit your ideas, as squeezing in too many can dilute the message. Ideally, your presentation will have one main premise that you'll support with different examples.

5. *Kill the complexity*: Talking about something complex? Use analogies to make abstract concepts tangible, especially when it comes to numbers. For instance, saying you need to analyze a petabyte of data is less impactful than explaining that a petabyte is the equivalent of taking four thousand digital photos every day for your entire life. In addition to simplifying concepts, simplify your language and avoid industry jargon. While some believe using big words will make them seem smarter, it typically makes the audience feel dumb—and you can't win them over if they don't understand you!

6. *Practice—a lot*: It's a myth that some people are natural-born public speakers. The confidence and charisma of expert presenters is less a gift and more a skill that's honed with time and repetition. The good news is that anyone, including you, can have it too.

For more information about storytelling, see Part V, *Data Alone Does Not Create Empathy—Storytelling Is Key*, page 155.

Understand and Speak the Language of Business

Dwayne Hill

In life, exhibiting the ability to speak and communicate effectively through language can enable enriching experiences, build trust, and reduce embarrassing moments. Effective speech and communication also carry over into the business world. Designers who can speak the language of business helps them communicate design decisions in narratives that resonate with business decision makers.

UX designers not only advocate for users; we also solve business problems. Whether you're a consultant or you work inside an organization, you're paid to solve a problem for the end user on behalf of a business that wants to increase profits, reduce expenses, and gain a competitive advantage.

The following terms will help you communicate with your business peers. These are just a small fraction of the terms you could learn, but knowing and using these will help you build positive relationships with your business partners since you'll be speaking from a perspective they understand.

- *Return on Investment (ROI)*: A way to measure the performance and efficiency of an investment. When companies invest in something, such as people, research, or product development, they want the benefits of their investment to outperform what the investment cost. If a company invests $500K a year (cost) in your team, then your team should generate more revenue than what it costs to employ you. Other ways of delivering ROI are to decrease company expenses by creating more efficient internal processes or by removing customer frustrations.

- *Profit*: A financial gain usually indicated by the difference between the amount earned and the amount spent buying, operating, or producing something. Profit measures the performance of the business. People are often willing to pay more for better materials or design, which can result in charging more for quality products and increasing profits (we see this, for example, with companies like Nike and Apple). Understanding how your design decisions can or will deliver profit helps get buy-in from business stakeholders because you're taking their perspective into consideration.

- *Revenue:* The total income generated or brought into a company from its primary business of selling a product or service.

- *Revenue model:* A framework that describes how a company generates revenue. It contains key components of a company's business model, identifying which revenue source to pursue, what value to offer, how to price the value, and who pays for the value. A few examples of revenue models are subscriptions (Netflix), freemiums (LinkedIn), and franchising (McDonald's). Knowing the revenue models of the company will help you understand how it makes money and provide insight on design direction.

- *Business model:* Describes the way a company generates value for its customers. A business model helps explain the overall strategy of a company. A few examples of business models are retail (brick and mortar), gig economies, and ecommerce. It's a good practice to tie your design decisions back to how you're generating value for the customer.

- *Profit margin:* How much money is left over from selling a product or service after deducting the cost of producing and selling that product or service. As a designer, if you can help decrease the amount it costs a company to produce a product or service, you can help increase the profit margin.

- *Trade-offs:* Decisions a company will make to diminish something in order to increase something else. Each day your business partners make several decisions. If you can help them make wiser decisions or alleviate decision fatigue, then you will be viewed as an asset.

As mentioned earlier, these terms are but a small sampling of business language we may encounter as designers, but knowing them will help us better explain our decisions to a business audience.

Expand Your Network Through Community Involvement

Jen McGinn

To go fast, go alone. To go far, go together.
—from Simon Sinek's book, *Together Is Better* (Penguin)

There are many paths to success and fulfillment as a user experience professional, but few, if any, involve working alone. As such, professional networking and community involvement will help boost your career. *Networking* is no longer simply the process of going out and introducing yourself; in our connected world, new connections will sometimes come to you.

Community involvement can be many things: attending a monthly meetup, volunteering at or attending a conference, giving a presentation, talking on a panel, writing an article for a UX magazine, teaching a class, volunteering your services, mentoring someone, or publishing a peer-reviewed paper. Really, it's anything UX-related that you're not getting paid for, or not getting paid much for. Sometimes you benefit; sometimes someone else does. Ideally, everyone benefits.

The connection between community involvement and professional networking is that community involvement allows you to create a broader network by meeting new people at events. Conferences are one of the best places to meet people outside your existing circles, because you are all there to learn about similar things. At meals, you can easily sit at a table where you don't know anyone and strike up a conversation about conference topics; or if you're there with a group and you see someone looking for a table, you can invite them to join you. Likewise, cocktail hours can be opportunities to join groups or conversations that you wouldn't otherwise have access to.

You can break the ice with some easy questions like "What do you do?" and "Where do you work?" Sometimes this leads to an awkward response like

"I'm not working right now," or "I'm looking for my first job in UX." At that point, you can say, "Tell me more about what you're looking for—what kinds of roles are you interested in, and at what kinds of companies?" You can inquire as to why and how they got into UX, what they did before, and what UX topics interest them. Everyone has a story to share; you just have to ask.

Another way to get connected to a broader community is through email lists and LinkedIn groups. The UTEST email list has been active for more than 20 years and is a safe space to ask questions. You can search for LinkedIn groups that interest you, and you can find people you admire on LinkedIn and ask if they will give you an informational interview or review your portfolio. This kind of request happens more often than you would think. As a UX professional, you can offer to review portfolios or be a mentor on LinkedIn, either as a free service or in return for receiving feedback on your portfolio.

You can also expand your current circle by hosting an in-person or virtual meetup, brunch, dinner, or cocktail hour—just ask each of your friends to "bring" someone you don't know. You can deepen the role you have within an existing circle of people by volunteering to redesign the website for a group you already belong to. One last suggestion: if you have a great idea for a conference presentation, panel, or paper, include your friends and colleagues. You'll end up with a better result, and when one of them has a great idea for a conference proposal or meetup talk, they'll invite you.

Amplify Your Value by Finding Advocates Outside Your Team

Catherine Dubut

As you progress in your UX career, it's imperative that you hone the skill of communicating your work's value in the context of the company's business. We already practice the day-in, day-out task of aligning with product management, development, and other partners. Beyond sharing designs and ideas with immediate stakeholders, it is crucial to find colleagues who can amplify your value. These people that you build relationships with can become your greatest advocates and can help share your work's value with the broader business.

Why search for such advocates? Without them, key information about the business challenges you are solving with design can inadvertently be left out of the conversation. When I overlooked developing these relationships, I became isolated in my thought patterns and was limited to the similar perspectives of those on my immediate team. And when my product team neglected to share our process beyond immediate stakeholders, our projects lacked visibility and ended up lower in business priority.

Finding these advocates can strengthen design's influence in the product development process. At a prior company, our director of product saw my work result in tangible outcomes, and he was familiar with my overall design process. The company considered outsourcing development work for a quick feature redesign, which I disagreed with. While presenting the proposed end-to-end journey to stakeholders, I illustrated the complexity of our outdated technology stack, highlighting risks with the design and development collaboration. The director quickly stepped in to change the work's direction, favoring in-house collaboration instead. His support resulted in a successful launch, positively impacting the business and strengthening our brand with customers. His advocacy of my work played a key role in my career at the company, while also contributing to the company's success.

Tapping into genuine curiosity while utilizing user research tactics can help you find and nurture these relationships:

- *Look at the organization chart:* Relative to where you sit, which leaders are most influential? Whose work are you most curious about? Which groups have goals that overlap with those of your team? These colleagues-turned-advocates can come from many disciplines—from marketing to merchandising and customer support.

- *Meet a potential advocate on a human level:* During a casual conversation over lunch or a simple snack date, ask questions like:

  — Can you explain what you do?

  — What are your team's biggest challenges?

  — How do your line of business and user experience relate to each other?

  — How can we partner together to make a bigger impact for our customers?

- *Connect and build:* Maintain a list of people you've met and track these growing relationships. Be present at cross-discipline events and team gatherings. You are not a name on a presentation deck! Your visibility enables others to see you as the impactful person behind your valuable work. You can also learn from the challenges your counterparts are facing within their disciplines. They may not understand aspects of the human-centered design process, but their knowing what you do and who you are will increase your chances of getting called on for visible opportunities.

These advocates are extra pairs of eyes who can validate how your design work relates to the business. With their perspective, you can extrapolate insightful connections that others on your team don't see. And most excitingly, these relationships can result in sponsorship, in which these leaders advocate for higher visibility projects that open career doors, a practice particularly important for underrepresented minorities.

Every human interaction can make our world feel a bit smaller and more connected, especially within large organizations, where individual faces and names are easily lost in communication. We should seek to apply this growth mindset to our work relationships as we rise in leadership.

Design Mentorship Is a Lifelong Commitment

Kristian Delacruz

You're a successful UX practitioner who has a few years under your belt. You've delivered features, led research efforts, and provided value to your team and company. After all of your work, you're probably evaluating the next step in your career. Mentoring can be a fulfilling and valuable next step. Although it's not for everyone, the benefits can ground you as a UX practitioner and sharpen your leadership skills, empathy, and shared values. Before diving into mentorship for the first time, consider some key moments you will face.

Remember that Past Experience Is Valuable

The design community thrives on people with diverse backgrounds and previous work experience. You might mentor an individual graduating from a UX boot camp with no design background, a fellow designer on your team, or a college graduate. To help your mentees, leverage their past experiences by encouraging them to find analogous past skills and determine how these can be applicable in design. For example, if they have past research skills, point them to ways they can increase their user research methodology knowledge. To help your mentees, consider the following tips:

- Precisely identify their key strengths and weaknesses.
- Guide them toward expanding their network to gather different perspectives on their skills.

Be a Cheerleader

When conditions get challenging for your mentees, find ways to continuously lift them up through the tough moments. Let's say a mentee told you they got to the last stage of an interview and wasn't accepted—be there for them, and motivate them to push forward. Your mentee is relying on you for guidance. Consider the following tips:

- Remember that feedback and critique are gifts.
- Be realistically optimistic.
- Block time for reflection.

Build a Lifelong Relationship

Be there during your mentees' successes and hardships; build a lifelong bond by maintaining an open line of communication. There may be periods of time when you don't speak with a mentee, but that's okay—just be sure to get back in touch at some point. For example, I had a mentee in 2018 who landed her first UX position. Later, when she had earned a new position at a software company, I made sure to get in touch and congratulate her. When you have mentees, be there for their monumental moments to build transparent trust.

Mentoring isn't easy. It requires patience, empathetic listening skills, and selflessness. Remember these two key final themes to make your mentoring style relatable.

Reach Out First

Proactively reaching out and letting your mentees know they're doing a great job gives them a signal that you care about their growth. Make this conversation genuine by including snippets of the work that the mentee accomplished, as well as feedback from others, if available.

Be Human

You're not perfect. You're as human as the person you're mentoring. Don't forget that you're also learning and facing daily challenges—professionally and personally. Mentoring gives you the opportunity to take a step back and to potentially see your own challenges in a new light. During a mentoring call, I met an individual who had recently finished a boot camp. I could tell she was losing motivation and lacked energy. The call pivoted from a portfolio review to a session to rediscover why she completed the boot camp. When you're mentoring, find the root cause of the mentee's problem using interview techniques, and remind yourself of some key principles:

- Always be vulnerable.
- Always be supportive.

Mentoring is a rewarding experience. With time and practice, you'll get better at it. Don't forget to reach out to your mentees and have a genuine conversation—we all deserve a lifelong connection in our lives.

Create a Design Portfolio that Gets Results

Shanae Chapman

A design portfolio is one of the most important tools for job seekers. How you display your portfolio and choose the content to include can determine whether or not you hear back about your dream job. Follow this formula for building a portfolio that lands you interviews. Anyone from entry-level designers to senior-level designers and beyond can use this strategy to showcase their work.

1. *Select a digital platform for your portfolio that is easy for you to use:* Consider selecting a no-code or low-code template as a starting point. You can also customize the template using the platform's editing features and make it represent your aesthetic as a designer. You can adjust the placement and dimensions of elements, fonts, icons, backgrounds, and more. My favorite framework for design portfolios is the WordPress Divi theme. Also consider Squarespace and Webflow, both of which have some portfolio templates available.

2. *Tell your story:* What drew you to design? Be creative in sharing your passion, personality, and career trajectory with personal photos, awards, favorite quotes, and efforts at community involvement that show who you are as a designer. Include any metrics you have regarding your design career, such as the number of user interviews you've conducted or the number of clients you've supported. Take the time to answer that familiar interview question "Tell me about yourself," but do it in a fun and creative way. You might also consider adding a short (one-minute) video reel to introduce yourself to portfolio viewers.

3. *Build a portfolio focused on your career aspirations and the type of role that interests you:* Make sure to get permission from your employer before posting portfolio pieces publicly. If you want to land a role with a strong research focus, plan on highlighting your research philosophy, your qualitative and quantitative skills, and case studies that show

positive results stemming from your work. If you want to focus on a role with a mix of research and interaction design, focus your portfolio on showing the end-to-end process from research to design. If you want to focus on visual design for your next role, focus your portfolio on projects that highlight design systems and modernizing visuals in legacy apps. If you want to focus on design leadership, focus your portfolio on examples of when you empowered team members, built teams, created supportive communication channels, and streamlined processes.

4. *Include an overview of your projects from start to finish*: For each project, share the problem and challenges experienced by the users. Highlight your solution and how it addresses the aforementioned issues. Share the actions you took by going through your iterative process step by step. Dive deep into how you solve problems and focus on project results. Share a mix of visuals and text to describe how you take a design problem and break it down into actionable steps to get to the agreed-upon solution. Share your research of industry benchmarks, user interview plans, usability testing notes, design studio sketches, low-fidelity wireframes, and high-fidelity mock-ups and prototypes. Also share the results of your work and any possible metrics such as increased conversion rates or decreased support calls.

5. *Make it easy for others to connect with you*: Provide a link to your resume highlighting your work experience, academic accomplishments, and design community involvement. For safety reasons, if you are posting your resume on your portfolio, consider listing your city and state instead of your full address. In addition, keep your LinkedIn profile up to date with your work experience, freelance projects, professional organizations, and industry events.

My final word of advice is that you should share your story with authenticity. Treat your portfolio like any other design project, get feedback on it, iterate, and evolve it over time. Highlight your strengths in building empathy for users and creatively solving problems through the power of design. Your unique perspective is an edge. Make recruiters excited to get to know the imaginative, thoughtful, and brilliant person you are.

PART II

Strategy

User Experience Extends Beyond the Digital Realm

Frances Close

If you survey the landscape of user experience classes offered online, you are likely to find that the majority focus on creating digital experiences. This is not surprising, as many of the capabilities of a UX designer are associated with the interactions between people and the *digital* products they use. However, in a world that feels more digital by the day, it's important to consider the human interactions that surround our digital ones and how the two types of interactions work together under the umbrella of UX.

An experience with a company consists of many touchpoints. First, someone learns about and interacts with a brand. Foundational brand elements like visual language and voice and tone set the stage for things to come. A flexible brand system that spans digital and physical spaces and speaks to customers at different points in their journey helps ensure a cohesive end-to-end experience. Next, as someone becomes more familiar with a brand, they engage in different ways. This might include buying a physical product they can open and touch, walking into an environment with a certain smell, or hearing a sound while feeling an indication via haptic feedback. While the last item in that list likely has a digital component, the other two are rooted in the physical world. At each of these points in the experience, people make natural connections between themselves and the brand, moving between physical and digital spaces and interacting with other people.

If we focus on and map full customer journeys and consider the everyday human interactions happening throughout, it opens us up to thinking beyond screens. In return, we can use learnings from the varying interactions across a journey to amplify brand presence and, more importantly, create meaningful connections with customers across entire ecosystems.

As a quick example, consider the process of using your phone to make a purchase. While the main transaction is digital, there are opportunities leading up to that moment that make it feel seamless and delightful. How does the interaction compare to the traditional model of paying with credit or cash? What can we glean, mimic, or avoid from the physical experience and infuse into our digital experience? What other actions could be enhanced by physical product design or information design?

To help you think beyond screens and consider how different human senses play into a digital experience, here are some things to consider:

- *You are designing for people:* What are those people doing, thinking, and feeling outside of your digital interface?

- *You are talking to people:* Leveraging a meaningful, clear, and consistent voice and tone is key.

- *People are drawn to tactile things:* Packaging, physical product design, and print materials play into a brand experience.

- *Space is experienced in a visceral way:* Surroundings that meet and appeal to the needs of customers can be a huge differentiator. Consider what people will see, smell, and feel.

- *Timing matters:* Short attention spans are real, so craft critical moments within the experience accordingly.

- *People crave human interaction:* Remember that you can't satisfy all needs through a screen. What physical touchpoint might you add to a digital experience to meet the need for that type of interaction?

As you approach your next UX project, step back and think holistically. If there are gaps you can't fill digitally that will enhance the overall experience—such as packaging design, brand design, or sound design—don't dismiss them simply because they fall outside the traditional UX wheelhouse. Think about ways to engage others who might be able to offer their expertise, and in the end, help create a complete customer experience that connects with people beyond their screens.

Know the Difference Between Experience Mapping and Journey Mapping

Darren Hood

Have you ever been curious about experience mapping and journey mapping? Let's take a quick look at some of their key factors, distinctions, and benefits.

Let's Level Set

To ensure we're communicating properly, let's level set by covering the terminology. The term *journey* refers to experiences and touchpoints from a holistic perspective. We would look to create a journey map when we seek to understand experiences and touchpoints in a broad sense. This approach provides a full-blown set of references that can highlight pain points, opportunities for improvement, system engagement, and sentiments. It can also foster needed discussion.

The term *experience* is focused on specific, extracted touchpoints and factors across an experience. While it can contain many of the same elements as a journey map, an experience map is not as broad (for example, it might focus on five to seven specific aspects of the experience). Also, because of its absence of breadth, an experience map can be produced in far less time than a journey map. For this reason, the creation of experience maps is far more popular. Time constraints (such as agile sprint lengths) may also limit your practice of and affinity to experience maps.

Alignment and Challenges

Now let's look at some of the similarities:

- Both types of maps should be based on actual data.
- Both help identify users and customers.
- Both provide ways to digest experiences visually.
- Both examine touchpoints and specific focal points.
- Both provide opportunities to drive greater understanding of needs.
- Both present sentiments and verbatims.
- Both serve as reference points that optimize project-related cognition.
- Both help us to identify as well as evaluate microexperiences and microinteractions.

Many people and SaaS resources use the terms *experience* and *journey* interchangeably. When exposed to the concepts of experience and journey maps for the first time, new practitioners practitioners may unknowingly experience anchoring bias—that is, they may assume the first presentation they are exposed to is accurate. When anchoring bias is in effect, someone hearing an accurate presentation regarding the two mapping methods will be confused. To overcome the associated ambiguity, we must strive for accuracy at the practical and cognitive levels, as this can help mitigate any issues at hand.

Takeaways and Reminders

Experience and journey maps can both be of great value to a UX practice. We simply need to understand the differences between the two and when each is strategically optimal. Consider these final takeaways:

- If you have time, consider creating a journey map. This way, you can look at everything—for example, your entire website or mobile app, examining everything from a login to the registration process to the way people review products and services to an ecommerce experience, leaving nothing out.
- If you lack time but would benefit from breaking down experiences, consider an experience map. For example, if your data shows high levels of abandonment during the checkout process in your site's ecommerce experience, you could limit your focus to each of that process's steps and touchpoints.

- If you don't have time to create a journey map as a dedicated project, you could still consider creating one on the side. This will help provide value for your team. It just wouldn't be directly connected to any project timelines or milestones.
- One of the greatest benefits of either mapping approach is using it to identify microexperiences (and we cannot attend to microexperiences if we do not identify them).

If mapping is not currently included in your UX toolbox, I hope I've provided you with enough food for thought to at least put it on your radar.

Design Customer Experiences, Not Features

Gail Giacobbe

Many UX designers have had the experience of designing a new feature and seeing it perform well in early user research, only for it to fall flat when it goes to market. One key reason this can happen is that designers may fall victim to wearing *feature blinders*: looking narrowly at a particular feature and failing to step back and look at the end-to-end customer experience.

I fell into the feature blinders trap while leading a project to improve the renewal experience for a subscription product. Our goal was to increase renewal rates, and our user research uncovered quite a few design flaws in the original renewals page. So we updated the page design, ran early user research, and refined the design until it performed well with our target audience. We then rolled it out broadly, confident the updated design would result in big improvements in renewal rates. But as we monitored the data, we observed only a small improvement—nothing near the rate of improvement we had been aiming for. What had happened?

We dug into the data and noticed that a large percentage of customers who failed to renew had an expired credit card on file. When we looked at renewal rates for that set of customers, we saw they failed to renew almost 100% of the time. Then it struck us—we had done nothing in our renewals page redesign to address expired credit cards! We had been working only to improve the renewals feature, when we should have been thinking about the overall customer experience.

So we zoomed out and looked at the entire customer journey. This broad, end-to-end view uncovered multiple ways to improve the renewals experience for all customers, including those with expiring credit cards: we worked with the email marketing team to send proactive emails several months before the expiration; we created a prominent warning to show in the interface when the expiration date was approaching; and we modernized the "update credit card" flow to make it quick and easy. With these improvements in place, we saw the desired increase in renewal rates.

Take off your feature blinders and tackle the full end-to-end customer experience by keeping a few things in mind:

- Use data to identify cohorts of customers with clusters of problems. For example, once we identified the common patterns related to credit card expiration, it was much easier to figure out where to focus design efforts to improve the customer experience and increase renewals.

- Once you know where the problems are for a specific set of customers, walk in the customer's shoes across their end-to-end experience. Take the time to use the product like a customer would, running through some real-world scenarios from start to finish.

- Examine all the touchpoints at which customers interact with your product or service, including search results, marketing communications, interactions with customer support, and product settings.

- Ask users what other channels or methods they use to accomplish the same goal. Find ways to make it easier for the customer to achieve their goals across multiple touchpoints.

By taking off your feature blinders and zooming out to understand the end-to-end customer experience, you can walk in your customers' shoes and map all the places across the journey at which the experience could be better. This journey map can then inform your delightful and smooth end-to-end customer experiences. To learn more about journey mapping and other strategic deliverables, check out the previous chapter.

Create a Truly Visible UX Team

Sonia V. Weaver

Despite the UX discipline existing for decades (under one name or another), companies still employ folks who are completely unaware of the UX team or the role it plays within their business. Sometimes, developers even get praised for UX's work. Don't get me wrong—we love our engineering colleagues—but they don't deliver all those intuitive user experiences by themselves. So ask yourself: are people in your organization aware of the UX team's existence and its impact on customer loyalty, user satisfaction, and the company's own bottom line?

If the answer is no, try the following techniques to help market your initiatives and successes. Since employing each of these over the past two years at my 1,500-person company, our UX team's visibility has gone from near zero to eleven (scale à la Spinal Tap). We sparked a company-wide discourse around usability and now receive extra focus and resources to meet our goals.

Suggested tactical techniques:

- *Create real-time conversation channels:* If your company uses Slack, Microsoft Teams, or some other tool that supports discussion channels by topic, create one or two channels dedicated to UX design and research. Advertise them using other channels, emails, company newsletters, and intranet posts. Don't just share information; use polls and questions to engage members.

- *Add content to internal company newsletters:* Leverage existing communications to spread awareness of your team's work. Get in touch with newsletter owners and offer to submit occasional content about important design or research projects.

- *Write a design- or research-related blog post on your company's external website or community portal:* Advertise it internally in the two spots mentioned previously. Work on the topic with your UX manager/director to get buy-in.

Suggested strategic techniques:

- *Create a quarterly UX newsletter and distribute it company-wide:* Set up a template for the type and number of articles, and repeat that with each issue. Keep articles to one page and use plain language so they're easy to read. UX team members can author articles that offer sneak peaks at upcoming designs, share research findings, or speak to topics like "design thinking." Be creative and avoid the dry, technical content that people might expect, and you'll be surprised how well it works.

- *Host an annual usability survey targeting hands-on operators:* Use quantitative metrics (SUS, UMUX-Lite), sentiment questions, and open feedback in a single survey covering one or more products. The results will interest (dismay? validate?) a wide number of internal teams. Present the findings to engineering and product management first, and then more widely. This event primes large numbers of people to think about usability at the same time.

- *Claim space on the main stage:* Often UX is nested under other disciplines, but that doesn't mean others should speak for us. UX leaders should advocate for time on All Hands or Town Hall calls, present at company conferences, and host content at user conferences. Find ways to claim some limelight. The board of directors, the CEO and other C-suite members, and employees across the company all benefit from understanding our mission, actions, and impact.

UX teams do fantastic work but often remain cloaked under some strange shadow. Or perhaps it's more like the movie *The Sixth Sense*, where only the special few can "see UX people." It doesn't have to be this way: leverage this list or use it to spark your own ideas. Drive conversations far and wide about the role that user-centered design (and your team) plays in your company's success. You'll be thankful when colleagues from outside your inner circle suddenly see you and recognize the value of your work.

Thinking About the Future Is Important for Any Design Process

Liz Possee Corthell

The future is something that's always on our minds, yet it's impossible to predict where the world is headed. This is one of the core challenges of strategy work: we can't predict the future, but we still have to act. Instead of predicting, we have to anticipate where the world is headed and ensure we create the future we want to live in.

Futures thinking is a method in which we use visions of the future to change the present. Most look to the past to measure success, not the future, but the world around us is getting more complex, so the past isn't useful anymore. As designers, we can apply futures thinking as a mindset to ensure we look beyond the horizon to consider our impact 10–15+ years into the future.

Futures thinking is a world rich with content and knowledge, and the complexities cannot adequately be summarized in one chapter. But there are methods we can use to begin designing the futures of tomorrow. Jim Dator, a professor and director of the University of Hawaii's Futures Studies Department, holds that there is no singular future:

> Futures studies is not about correctly predicting The Future. It is about understanding the varieties and sources of different images of the future, and of coming to see that futures studies does not study "the future," but rather, among other things, studies "images of the future."[1]

So how do we create "images of the future"?

First, we have to ask the right question. I recently worked on a project in which we asked what the future of work might look like, following the shifts

---

1 Jim Dator, "Alternative Futures at the Manoa School," Journal of Futures Studies 14, no. 2 (2009): 1–18. https://oreil.ly/HiZCn.

in how we work and live amid the COVID-19 pandemic. To consider this, we had to understand several factors:

1. *The past:* What is the history?
2. *The near future:* What are the current forecasts and trends?
3. *Our assumptions:* What assumptions have we made as a team?
4. *The alternative futures:* Based on those assumptions, what alternative future scenarios can we create?
5. *Our goals:* What is the preferred future?
6. *Our path forward:* How do we get there?

As we grew to understand how office work had changed rapidly from an in-person model to primarily being remote, we saw emerging trends taking place in real time. Tech companies began announcing they would encourage their employees to be remote forever, which led us to assume that other industries might follow suit. This helped us create alternative future scenarios, including:

- *Remote-only:* Employees continue to work remotely forever, invest more money into their home offices, and move to states farther away from their offices.
- *In-person first:* Employees return en masse to offices, with a few exceptions, and continue to commute from their nearby homes when it's safe to do so.
- *Hybrid:* Some employees return, while others remain remote, but it is each employee's choice based on their role and responsibilities.

After presenting these concepts to stakeholders, we understood that the preferred future is hybrid. In exploring how to get there, we realized that a hybrid workforce has to be well designed for all ways of working. We explored some ways of doing this, including creating "phigital" spaces that mix technology and space with tools like VR headsets and virtual whiteboards.

Though a future of work that includes a VR headset may feel ridiculous today, as Dator says, "Any useful statement about the future should appear to be ridiculous." It takes courage to consider the future as designers, but it's something we all must do today to create the world we want to live in tomorrow.

Implement Service Design in Your Practice

Eduardo Ortiz

During your career, you will probably work on a service, whether a new service or an existing one that you're improving. While in either scenario the work to be performed is based on the same principles of design you've learned, the approach, scope, and stakeholders (including end users) are varied and different.

Let's start by developing a common understanding: *service design* is the practice of identifying infrastructure, planning processes, organizing people, and implementing solutions around a service. This is the basic structure of the phases of service design:

- Research (infrastructure, policies, people)
- Plan (organize and catalog the research findings, identify pain points)
- Apply (document potential solutions into a road map, note requirements or challenges)

Research

Research is the foundation of any (service) design project; it helps us develop the knowledge necessary for us to be trusted partners in solving the identified problem. It is through research that we validate the problem and identify the different stakeholders: business owners, customers, operations personnel—anyone affected by the service. We also learn about existing infrastructure, future planned upgrades, and existing policies or rules and who they apply to.

Research provides us with a common language through which to view the challenges we're working on. Common activities include the following:

- Interviews (1:1)
- Listening/observation sessions
- Desk research (reading literature)

How do we achieve this phase?

- Ask questions (interview).
- Get to the most granular problem statement possible (problem statement definition).
- Group and sort the findings from the questions you've asked (synthesis).
- Identify the different types of people to whom you've spoken or who have been brought to your attention (stakeholders and users).

Plan

You now have a plethora of information based on your research, which means you can start visualizing what you've found against the service and overlay the problems identified along the way (service blueprint). These visualizations not only enable others outside your team to develop their own deep understanding of the service and its challenges, but they also point to potential misunderstandings and areas that your research may not have uncovered (research is continuous). The planning phase enables you to bring others along and to learn from each other. Common activities include:

- Content/findings grouping (affinity mapping)
- Journey maps (visualize findings)
- Ideation sessions (you have problems, so work to identify solutions)

How do we achieve this phase?

- Use sticky notes or an online tool to share findings and catalog them.
- Group information by topics important to the service you're working on.
- Place these groupings across a timeline of the service, including the people affected; visualize the journey.
- Share your work and collect feedback to improve it.

Apply

Now that you've grouped your findings, identified requirements to improve them, and validated the research, you need to start laying out these solutions in a manner in which they can be executed. This means prioritizing the work you did in the planning phase while accounting for potential impact, costs, and feasibility of the solution identified. Common activities include:

- Prioritization workshops (organize information based on cost, complexity, what's important to stakeholders, and so forth)
- Service blueprints (you're now iterating on the journey maps based on the prioritization workshop outcomes)
- Readout sessions (share what you've learned with others, gather feedback)
- Road map creation (apply a timeline to the work identified)

How do we achieve this phase?

- Host stakeholder workshops to review findings and solutions (and determine what it would take to achieve them, how important these solutions are to them, how much it would cost to apply the solution, and so on).
- Revisit the journey maps and augment them with the different infrastructure touchpoints and policy intersections; they will now become service blueprints.
- Share your work and collect feedback.

These phases and approaches should serve as a baseline for implementing service design. However, the underlying theme is that it takes a team to achieve the desired goals.

PART III

Design

Don't Forget About Information Architecture

Joe Sokohl

A few years back, my wife and I wanted to clean up our utterly disarrayed basement. Our neighbor, a professional organizer, helped us organize 30 years of accumulation. Her succinct process of physical organization reminded me what information architecture truly is.

"Put like with like. Decide what to retain versus what to trash or donate. Then label and store items based on how you live," she said.

That's what information architecture (IA) does: *it organizes and classifies digital stuff so people can find and use it.* Though IA has existed as a field for well over 20 years, lately it has taken a back seat to other aspects of UX such as UI, content strategy, and interaction design. However, IA continues to be an important piece of the UX puzzle.

Like all other aspects of UX, IA starts with understanding the user and their context. User research resides at the core of IA. As we look at how people use information, we observe their behavior, both in seeking information and in storing it. How people engage with information is similar to how humans have engaged with their environment for eons. "Information berry picking" illustrates how a person goes from one information repository to another, finding information that helps them form a sense of meaning or place. "Information foraging" extends the metaphor, indicating that people look for those areas of information that pose potential value. People wander digital aisles as well, looking for meaning in the informational hallways they walk.

To organize information, you first have to *know what information you want to organize.* Is it existing information that needs better arrangement? Is there information that needs creating? Is there duplicate information that needs reconciling? To help us put like with like while we determine what needs to be retained, bundled together, or discarded, we use:

- *Content inventory:* Identifying every piece of content in the information space
- *Content audit:* Analyzing the content by type, purpose, redundancy, and viability to the user

To understand how people name the stuff they use, we conduct:

- *Search log file analysis:* Reviewing terms that users entered when searching for things
- *Card sorting:* Letting users sort terms into categories that make sense to them
- *Tree testing:* Asking people where they go to find or do something

With a solid terminology in place, we label information spaces, finding ways to store them for users to access and use. Once they are labeled, we then provide guideposts and markers to help users wander through information spaces to find what they're looking for.

Jorge Arango captured IA so well when he talked about IA's concern "with the structural integrity of meaning across contexts." IA helps people engage with that meaning across all experience touchpoints.

Site maps, taxonomies, and search results pages don't provide enough framework for emerging interactions. Voice and virtual and augmented experiences all need structure and classification, just as visual ones do. For example, voice interfaces require IA. Providing information clues reduces the memory people must retain. A clear aural framework helps people know where they are when asking for an album to play, a forecast to be delivered, or a recipe to cook.

Our main human concern in IA remains not just the structure but what people do within it. As Frank Lloyd Wright quotes Lao-Tzu at Taliesin West, "The reality of the building does not consist in roof and walls but in the space within to be lived in." In the same way, IA consists of the space in which people live their information lives. It provides the structure, the bones, and the framework for other activities to operate in.

For information about identifying IA problems, see Part V, *Know These Warning Signs of Information Architecture Problems*, page 147.

When Prototyping, Consider Both Visual Fidelity and Functional Fidelity

Chris Callaghan

Most UX design tools focus on visual elements such as typography, graphics, motion, and CSS handoff. Although many tools promise prototyping, few actually provide the features needed to adequately architect inputs, outputs, stateful components, or conditional elements. For example, many "UX design" tools don't provide the features to create a working form field, a core element of many experiences.

The promise of prototyping is broken, and designers are left without the necessary tools to explore and communicate the fundamental behaviors of user interfaces. This results in an interaction design blind spot.

I once encountered a perfect example of this blind spot as I prepared a usability study for an ecommerce client. My client wanted to test a new checkout their agency had designed. As you'd expect from a checkout, there were many key experiences, such as sign-up journeys, basket management, delivery selections, payment details, and order summaries.

The agency told me they'd provide a "high fidelity" prototype for me to usability test. Timeframes were tight, and as the study date drew near, I became increasingly concerned that the prototype had not yet been shared with me. But at the last minute, the agency delivered it.

However, there was a problem. Though the prototype was high in *visual* fidelity, it was low in *functional* fidelity. The prototype *looked* like a real checkout, with accurate typography, beautiful iconography, and button design, but it didn't *work* or *feel* like a real checkout.

Because the UI designer chose a popular visual design tool masquerading as a UX tool, the best they could do was to stitch their artboards together, resulting in a low functional fidelity prototype. This meant the user could navigate only in the specific order in which the screens were stitched. Because of this, study participants would try clicking in places that didn't work and ultimately would disengage within seconds.

A further problem with the artboard approach was that none of the editable elements of the interface were actually editable. The UX design tool of choice didn't offer much beyond surface design, so the UI designer was unable to design or prototype basket functionality, payment selectors, or the delivery details form. The designer had to generate vast sequences of artboards in which form filling and keyboard entry were simulated with dummy information. Again, study participants would disengage as they were presented with irrelevant placeholder information and experienced click-to-continue fatigue as they stepped through simulated interactions.

Fortunately, I had prior experience that enabled me to spot these concerns ahead of time and demonstrate them by way of a few pilot tests. The appreciative client rescheduled the study, and I produced a new, fully interactive prototype in a more appropriate tool with high visual *and* high functional fidelity.

For this project, a high functional fidelity prototype was critical to testing the experience, a fact demonstrated numerous times as we observed participants interact with stateful components at moments of their choosing without needing to suspend disbelief.

So I'd encourage you to think about the learnings you need to make from your prototypes and to consider not just the *level* of fidelity you need but also the *type* of fidelity you need. And remember: if you only surround yourself with surface design tools, you'll only ever design the surface.

See Beyond the "Average" User

Hillary Carey

As designers, leveraging standpoint theory can help us value, promote, and integrate the perspectives of diverse people and contexts. It says that people who navigate systems from the margins have a better understanding of them. They see the boundaries and are more aware of how things work because they bump up against the edges. Overlapping and compounding experiences of race, class, gender, nationality, religion, ability, and sexuality shape people's life experiences. Therefore, drawing on other perspectives helps to challenge what we otherwise take for granted.

Think about the systems you design. Who moves through them with the fewest barriers? Describe that person. Are they typically white, affluent, able-bodied, cisgender, and English-named? Does your system break when someone outside those norms tries to navigate it? In their book *Design Justice* (MIT Press), MIT professor Sasha Costanza-Chock describes these fictional personas as "unmarked users." Costanza-Chock describes the painful experience of being a transgender person navigating physical and digital systems that don't acknowledge their body. TSA screening systems and data systems that pull from historical records that misgender them force security agents to indicate "male" or "female" based on an initial glance. This inflexible interface forces the cognitive load onto the individual user rather than demanding more technology accommodation.

Examine the assumptions in your design work and acknowledge your contexts. What innovations might you uncover by examining the path of less typically centered users? Consider the App Store, the smartphone ecosystem, your business context, your location, and the economy your product or service exists within; then map it. Draw concentric or overlapping circles to identify systems that surround your design challenge. On which systems is your solution dependent on or part of? It can be powerful to situate your products within a larger context.

Now think about who sees these systems from a different point of view than your unmarked user and from yourself. It can be eye-opening to interview people who might hate your solution—those who don't need it and those who would deliberately create an alternative. It's the flip side of looking squarely at your average users. Average people, if they actually exist, don't challenge your thinking. Real innovation comes from tension and pushback; learn from people who don't easily navigate your systems.

Designers constantly make choices that shape which actions people will experience affordances and disaffordances. Affordances encourage particular actions, and disaffordances prevent specific actions. Costanza-Chock describes the small but ongoing othering that occurs in interfaces as *UX microaggressions*. Requiring more work of people who are not the unmarked user is one instance—for example, if "male" is always the default on your menus and other people have to make the effort to change the response. There is emotional labor as well in never seeing oneself as the norm the product was designed for. Taking on alternate points of view can help you uncover what and who is constrained or privileged throughout the design.

Asking three fundamental questions of your design can help you increase your awareness of a broader range of user experiences:

1. Who is prioritized?
2. Who is left out?
3. Who will benefit?

The concept of *standpoint*—that people who experience systems from the margins will have more insight into how they truly work—can introduce more accurate and justice-oriented perspectives into our work (for more, see Alison Wylie's "Why Standpoint Matters" (*https://oreil.ly/OXHzu*)). Embracing contextual awareness and then deliberately involving people with alternative standpoints can ensure you understand the opportunity from more than one angle. There is great wisdom to be found—it will extend our creative capacity if we deliberately incorporate diverse perspectives.

Work Together to Create Inclusive Products

Al Lopez

Our goal as user experience professionals is to create products that resonate with our audience of users. We never want to create a product that leaves a user feeling as if the product was not built for them. To create products that resonate with our audience of focus, we must have diversity of thought. Diversity of thought comes only from diverse teams. Part of our responsibility as UX professionals is to be advocates for diversity. It is the ethical thing to do. If ethics aren't reason enough, think about how global and connected our world has become. To keep up with today's global society, UX professionals and companies must push for diverse teams.

When people talk about *diversity*, what does that actually mean? Many people consider gender and ethnicity when thinking about diversity. But there are also age, religion, culture, sexual orientation, ability, and lived experience, just to name a few more categories. Inclusive teams contain people from various backgrounds, and teams with diverse backgrounds create more diverse solutions.

There are several ways UX professionals can include more diversity in our day-to-day practices. One way we can increase diversity is just by including actual diverse people in our studies and designs. Why can't a fictional persona be named Khadijah? If you are testing usability of a site, is there a reason you wouldn't test how the site works with people who use screen readers? Including diverse people in our designs and studies creates a solution that works for a larger variety of people. See Part V, *Diverse Participant Recruiting Is Critical to Authentic User Research*, page 143 for more information about recruiting a diverse population for your research.

To further increase diversity, if you are in a privileged group, stop giving job references only to friends who look, act, or think like you. I know this one is hard, but it's necessary. We share similarities with our friends that enable them to be our friends. Promoting only our friends for jobs takes opportunity away from more diverse applicants. If you and all of the friends you promote have college degrees, that leaves out UX professionals with boot camp backgrounds.

Another way to increase diversity is to remove bias from our work. We all come with bias, whether we want to admit it or not. But we don't have to live in that bias. We can and must think outside the box when it comes to our solutions. Are we constantly considering different ways of viewing the product? Are we getting diverse feedback on what we build?

It is not enough simply to have diversity; we must also create environments that allow for team diversity to thrive. To enable diverse ideas and perspectives, we must actually be open to diverse ideas and perspectives—one doesn't happen without the other.

We must also address any systemic issues that exist within companies and create safe environments for diversity to thrive. If you see something, speak up. This can be hard at times, but those of us in privileged positions owe this to our less privileged UX cousins. If you notice that only people of a certain skin tone are hired, ask why. If the trans person on the team is the only one who is consistently asked to work late, ask why. We must call out racism, sexism, homophobia, and so on, so we can create environments where those -isms do not exist.

All of us can assist in creating more inclusive products by increasing diversity. We owe it to our users and ourselves to create products that actually work for the real world we live in.

Advocate for Accessibility

Holly Schroeder

> Many seemingly small things significantly impact disabled folks. Accessibility, or lack thereof, is one of these things. It's as easy to make things accessible as it is not to. Implementing accessibility is your chance to have a real and positive impact on disabled people's lives. What's stopping you, then?
>
> —Nicolas Steenhout

Accessibility should be a part of our collective effort as designers. "But we have an accessibility expert!" Sorry, you don't get a pass. To be truly inclusive requires collective effort. Being an expert isn't required to add value, but we might have to step outside our comfort zone and dedicate time to learning.

Worldwide, one billion people are disabled, according to the World Health Organization. The internet provides abundant resources for learning the specifics of designing digital products to meet standards. However, accessibility and inclusion are more than just checklists. We should set aside time to ask ourselves the following questions and record the responses:

- When planning, are we thinking about including disabled people, and about where we may be excluding them?
- How are we personally ensuring accessibility is the standard?
- Are we dedicating time to learning how we can better understand the experiences and challenges of disabled people?
- Do we harbor bias?

Plan for personal and professional growth. Review the responses to the preceding questions and use the following prompts for further exploration:

- Evaluate projects and ensure disabled users are included.
- When disabled people are overlooked, speak up:
  - "What can we do as a group to make sure we include disabled people?"

— "What are our plans to test for accessibility?"

— Remind others it's a collective responsibility.

- Take steps to ensure a psychologically safe and inclusive space for disabled colleagues. Don't fall prey to the bystander effect and assume someone else has it covered.

- Don't rely on disabled people for education. Get familiar with the international standards for web accessibility (*https://webaim.org/*).

- Explore disability websites; there is publicly available information about nearly every disability.

- Consider how different disabilities may impact experiences and the barriers people may encounter.

- Attend webinars and meetups, listen to podcasts, and read about accessibility and how to be inclusive.

- Seek out autobiographical videos and blogs to better understand the viewpoints of disabled persons across the spectrum of disabilities.

- Leverage social media; most have disability communities—follow, listen, learn, and ask questions if it's appropriate.

- Reach out to disabled people. Ask if it would be okay to inquire about their disability to better understand their experiences and challenges.

- When asking questions, privately or publicly, get consent first. It's okay if the ask isn't perfect. Be open to feedback about how to craft future inquiries.

- Most people are willing to chat when asked. If they decline, don't take it personally. Being disabled is hard; sometimes we don't have the energy. Move on and try again.

- If a disabled person says a phrase or word is offensive, listen. Don't be defensive—it's devaluing.

- Pick apart biases. Abandon assumptions and old narratives—rely on facts.

- Practice observing your thoughts and behavior. Ask yourself why; ask five times to get at root causes. Don't waste time on shame; just try to do better next time.

As a practitioner in a field grounded in empathy, I know that taking the time to educate oneself and increase understanding is an investment that pays dividends. Accessible design and inclusive practices that develop into habits build stronger and more diverse teams, increase customer satisfaction and return on investment, and mean that less time is spent on rework. Perhaps most importantly, the investment is progress toward a global community that is useful and welcoming for all.

Design for Universal Usability

Ann Chadwick-Dias

When designing for universal usability, there are some easy and useful guidelines to ensure you are designing in a way that allows for the broadest range of users to use your design. These guidelines not only make your design more usable for people who may use various assistive technologies (ATs), but they also help your design integrate more seamlessly with different types of user input (voice, eye tracking, touch, sip and puff, search engines, AI bots, and so on). Universal usability improves the UX for ALL users, humans and ATs.

Here are five guidelines for universally usable designs:

1. *Keyboard access:* Ensure full keyboard access to every element in your design (typically via tab or arrow keys), and stipulate in the code the order in which elements should be represented, including actionable items and static content. Making your design fully keyboard accessible also makes it accessible to ATs—remember, many users cannot use a mouse.

2. *Color and contrast:* Use colors to complement the meaning of elements and to enhance your visual design. But never use color alone to convey meaning—for example, red for loss and green for gain in financial services; too many people have color blindness, which might make it impossible to see a difference, especially between red and green. Instead, use a redundant cue: - and +, in this case. Ensure important content has the highest contrast (75%+) and main content has adequate contrast (50%+). Contrast below 50% is difficult for many users to see but may be fine for some visual design elements. Avoid meaningful elements below 33% contrast—users can't easily see this.

3. *Images versus text:* Use images for images and text for text. To precisely control visual design text, designers often use images to represent text. But ATs cannot discern text in images, rendering them unreadable for

people who rely on the technology; use text when conveying textual content. The text should also be scalable and should allow users to magnify. Images used for visual design to enhance aesthetics, such as borders and 3D effects, should be "silenced" to ATs by using null (`alt=""`) for the Alt Text. When images convey actual meaning, always use concise and clear Alt Text, which conveys the image meaning.

4. *Headings:* Headings enable users to quickly scan and easily move their eyes from topic to topic. These shouldn't just *look* more prominent than the main text; they also need to be coded as Headings (H1, H2, and so on). Headings are a primary way that ATs navigate content.

5. *Text:* Use scalable/flexible text size and styles to make sure everything is readable; there should be a clear visual hierarchy of the IA represented in the code. Eye-tracking data confirms that users scan for prominent text. Research with blind users demonstrates the same thing—they navigate best when scanning the page for Headings in ATs. Design your text to both visually and semantically demonstrate a clear information hierarchy.

While there are many guidelines for making your designs inclusive, the point of universal usability is to provide an "equivalent" experience for all users, independent of ability. Whether they are neurodiverse, mobility impaired, blind, deaf, or any other range of abilities, they should be able to easily navigate and understand your design using any (or no) assistive technologies/devices. Sidewalk curb cuts are probably the most widely understood implementation of inclusive design; designed for people who use a wheelchair, they are widely used by many other users/devices (strollers, bicycles, and so forth). "Design for good" is just "good design" for everyone.

Inclusive Design Creates Products that Work for Everyone

Christopher S. LaRoche

Inclusive design is a methodology that requires including the full range of human diversity in all aspects and differentiations. It is the natural evolution and progression of both accessibility and universal design. Accessibility focuses on remediation of an application that creates a situation in which there is a perceived "mismatch" between the human interacting and the technology being used. Universal design emerges from physical and architectural environments, where the goal is creating a product for which one size fits all. Inclusive design recognizes that all humans have unique needs and qualities and formalizes creating designs that work for each individual through technological flexibility, so the product or application will be usable by everyone. This methodology focuses on digital environments, as they are more flexible than physical spaces.

The requirements of inclusive design avoid the exclusion of populations that is often accepted within technology and digital realms. For example, the discriminatory 80/20 rule is often used when creating software that justifies excluding part of the population. Using inclusive design in creating digital products results in a one-size-fits-one option for all that use the product.

The concepts of inclusive design were created by Jutta Treviranus, founder of the Inclusive Design Research Centre (*https://oreil.ly/ZNJvx*), and include three dimensions:

1. *Recognize and design for human uniqueness:* Fully understand the unique and diverse aspect of all people and their needs so that a monolithic solution or a separate "accessible" solution is avoided.

2. *Use inclusive tools and processes:* Design products through participatory design and co-design with populations traditionally not included in

design, including people with a variety of disabilities. Using tools that are usable and accessible to all is critical, as is open source technology.

3. *Design for a broad social impact:* Designing your product so it will have a wider cultural and societal benefit than a specific organizational focus is critical; the concept of using a ramp in lieu of stairs is a solid example of usefulness beyond original intentions.

The evolution and growth of inclusive design has expanded into technology organizations. Over the last few years, Microsoft has embraced and made inclusive design part of its overall product design and development. In fact, Microsoft has updated inclusive design aspects to fit its organizational needs and defines inclusive design according to these dimensions:

- Recognize exclusion
- Solve for one, extend to many
- Learn from diversity

The recent Xbox Adaptive Controller from Microsoft is an example of a product designed according to inclusive design principles. The concept of inclusive design has proliferated across many Microsoft products and services, so its products can be more useful and accessible to all.

Using inclusive design makes all products more usable and accessible. Think beyond making products for only the majority and always include designing for "edge cases" to make products more useful for everyone! Start today by reaching out to populations and groups that you have not included in previous research and design activities to create products that will include and work for *all* users.

Define What Your Design Does Not Do

Georgiy Chernyavsky

Design and functional specifications are awesome. Together with high-fidelity prototypes, they deliver the knowledge of how a product should be built. But ambiguity within and misinterpretation of these specifications can stand in the way of delivering great products by creating misalignment among designers, engineers, and other involved parties. There is one small trick that will help you avoid this: define what your designs *don't* do. We rightfully focus on explaining how our designs work; however, we rarely explicitly mention how they should *not* work. Including these points in your specifications will help bring clarity to important design decisions.

There are several ways to communicate these points: you can add them to your specifications document, create annotated screen designs, or simply mention them while presenting your work. Whichever approach you choose, defining functional points that your design doesn't cover, or even how the product should explicitly not behave, will help prevent situations in which the product is taken in an undesirable direction.

For example, imagine you're working on a new page for a radio show's website. People use this page to send birthday wishes for loved ones that DJs will read live during the show. Your team tests, polishes, and prepares the design for implementation.

The deliverables seem clear, and your team has few questions. Once deployed, you see the first metrics for your new page, and they don't look good: the bounce rate and rage click counts are high, as users seem to struggle with the form and rarely submit it.

After some testing, you determine the form has autocomplete enabled, just like every other form on the website. When users try to enter their loved ones' information, they get annoyed to see their own data filled into the form. In the end, the problem turns out to be a quick fix, but the experience of the early users suffered.

It can be dangerous to assume that your designs are self-explanatory—such assumptions often lead to miscommunication. To avoid this, make sure your design covers all necessary interface states, so your engineering colleagues don't need to invent those on the fly. Also, don't assume colleagues will find the desired interface states in your design library. Instead, make sure to proactively show them.

When you first sit down to define what your designs don't do, it may seem like a separate project. This is especially true if you're working on a complex product with dozens of features and hundreds of screens.

Focus on explaining the most complicated, innovative, and unconventional parts of your design. For the most complex interactions, make sure you thoroughly explain each user story.

Last, pick your delivery format. If you are working on a small team, mentioning the points about what your design doesn't do in a presentation or team session can be sufficient. But if you work in a large organization, where collaboration often happens in different time zones and sometimes stretches over months, detailed artifacts should be created so that they can be found and utilized by future designers.

Design documentation can be an exhausting task. However, if done right, it can help teams stay aligned, preventing time and budget losses. Defining what your design does not do can be a great practice; keep these simple points in mind to make your design specifications shine:

- Know existing design standards so you know when there is a need to deviate from them.
- Don't assume your design is "common sense"—documentation is the key.
- Show how the design should work and explain how it shouldn't.

Use Design Goals to Make Design Decisions Explainable and Defendable

Helmut Degen

Early in my career, I used the following design process: after finishing the research, I explored interaction concepts loosely connected to the research results. But I found it difficult to explain how the research results and project goals motivated the interaction concepts. Due to this, some stakeholders perceived the explanations as being made-up. In more than one case, a poor explanation or total lack of explanation led to a project stop.

To overcome this problem, I started using *design goals*. Design goals are derived from project goals and from research results. Here is an example: in one of my projects, I was asked to design an industrial engineering tool. Based on the research phase and the project goals, my team and I identified five design goals:

- DG.1: Single data entry (avoid redundant data entry)
- DG.2: Support reuse of project artifacts
- DG.3: Provide large canvas space
- DG.4: Enable comparison of project artifacts
- DG.5: Support adding new project templates

When we presented the design goals, they were backed up by research insights and project goals. It is critical that the design goals are easy for the project stakeholders to grasp: the design goal language should be understandable (avoid jargon), and the total number of design goals should be 10 or fewer. Once project stakeholders agree to the design goals, start exploring interaction concept options.

Design goals are the success factors that address user needs, business needs, and opportunities to differentiate the design from existing designs.

For the ideation phase, we relied on the design goals to explore, refine, and select an interaction concept. We initially explored four different interaction concepts (IC.1,..., IC.4) and then measured how they fulfilled the design goals; none of them met all of the goals. We then created a fifth interaction concept (IC.5) that addressed the limitations of the other four interaction concepts—it was designed to meet all five design goals.

When we presented the interaction concepts to the stakeholders, we first showed the interaction concepts IC.1 through IC.4, with their limitations, and afterwards showed IC.5. The stakeholders could easily follow the rationale behind the design goal–driven approach. They accepted the proposed interaction concept IC.5 without disagreement.

The identification of design goals starts when the research phase produces concerns and design directions. *Significant* design directions are selected as design goals. *Significant* means that the goal is critical for user success and costly to introduce late in the design and development process. For example, consider the design direction "make items comparable." Research shows that it is critical for success and requires the layout to display two sets of items at the same time, including highlighting differences. These requirements make it costly to introduce this goal later in the process. On the other hand, here's an example of an insignificant design goal: "use positive wording." Wording might be critical for the success, but it can be changed later at low cost.

Using design goals is a way to systematically explore, refine, and select interaction concepts or other design results. Design goals make design decisions *explainable* and *defendable*. They also help translate design decisions into a language that's more comprehensible for stakeholders. Finally, they can help show how user experience is a business-critical function with an evidence-driven design approach, contributing to the success of projects.

Think Synthetically to Design Systematically

Drew Condon

As UX practitioners, analytical thinking is so essential to our work we're often not even aware we're doing it. Analysis is a fantastic way to understand parts and pieces of a user's experience, but it's not as appropriate for understanding how those parts *work together* to deliver the experience.

When breaking experiences down, we can often lose sight of the interactions between the parts. As those interactions become less and less visible, our instinct can be to increase analysis of the pieces by breaking them down further. When we do this, we risk getting caught in a cycle of analysis. In trying to better understand the parts, we increase our distance from the whole. (See Gary Bartlett's "Systemic Thinking" paper (*https://oreil.ly/0C6Lp*).)

Sometimes we need help understanding interactions and how things work together. One approach is *synthesis* or *synthetic thinking*. Synthesis is a great tool for sensemaking; it provides us some clarity about how experiences work, or ought to work, together. Synthesis is about finding similarities and deliberately designing usable patterns across an experience.

UX practitioners need both ways of thinking, but the importance and utility of synthesis increases as the size and complexity of the things we work on grow and become more oriented toward systems.

Large technology platforms in particular rely on *shared flexible systems* to efficiently build experiences that suit diverse customer needs. For example, common experience systems of a customer resource management (CRM) system (contact records, workflows, communication interfaces) can be flexed and extended to support sales, marketing, and service customers equally well without creating three distinctly designed and engineered products.

To do this, we must uncover the similarities between needs to create the initial patterns. Dr. Russell Ackoff has a useful approach for how to apply synthetic thinking to systems in general. We can use his model for inspecting the experiences we create. Ask yourself these three questions as you approach designing something that's part of a complex system:

1. *What system is this experience a part of?* All systems are part of bigger systems. Where does the design you're creating sit within the universe of the other systems it touches? Try drafting an object model to help map it out.

   You're working on a ridesharing app. Map out the objects (ride, map, car, trip, and so on) and their functions and relationships to one another.

2. *How does that containing system work?* Every experience is affected by and affects its role in the larger system. If you improve only the experience you're working on, what is the impact on the experience above, below, and beside it? Where in the object model do relationships overlap or repeat?

   You're improving just the live ride-tracking experience. Does your improvement affect the pre-trip experience? The receipt experience?

3. *What key role does the experience you're working on play in the systems containing it?* Experiences have critical parts that can't be divided; this helps define the role they play in a containing system. What key role is the part you're working on playing in your systems? Can you divide any functions or have them be done elsewhere?

   How can you show position on a map? Does that system for showing position work in the other experiences that show a map?

Use these three questions to help make sense of the web of interactions and systems working together to deliver the overall experience. Find the patterns between them and reduce needless complexity and inconsistency where possible.

To do this well, we need to embrace the understanding that, as Ackoff asserts, "a system is never the sum of its parts; it's the product of their interaction."

Best and Last Impressions Are Lasting Impressions

Andrea Mancini

Think about the last time you ate ice cream on a cone. The ice cream was delicious; you clearly remember this detail. And after the ice cream comes the cone. What happens if the cone tastes like cardboard? You are more likely to remember this negative detail. But why?

We live daily experiences and preserve memories of these past moments, but we obviously can't remember every detail. Instead, we focus on strong emotions and feelings; we record "snapshots." But this is not the full story. There is a time-based factor that modifies how we remember experiences.

Think again about the ice-cream example. You had a negative lasting impression because you experienced the worst part at the end. Designers who are aware of this behavior can design more enjoyable experiences. If you can control how the experience is remembered, you can influence the user's final evaluation. This is the peak–end rule value.

This statement might sound odd since we often rely on retrospective evaluation to rate the quality of past experiences. That method doesn't necessarily give us the whole story, though. Two different mental processes are involved in the operation: memory recall and an act of evaluation, processes that are not completely logical. A famous 1993 psychological experiment conducted by Daniel Kahneman and Barbara Fredrickson shows a deep temporal component. Instead of remembering the full story, we mostly remember intense positive or negative moments (peaks) and the final moment of an experience (the end). These values are the most important ones in a retrospective evaluation.

This means that when we design interfaces and experiences, we must pay attention to the most intense/emotional moments in the customer journey but also must pay close attention to the final moments.

As a designer, you should aim to create positive peaks. A helpful illustration, good copy, or a nice design placed at the end of a positive interaction can help build a good impression in users' memories. By reinforcing positive peaks, you can help to also reinforce a favorable memory of the experience. You can also use this principle to boost engagement. For example, online courses are usually split into chapters to give learners intermediate achievements as a positive peak to motivate them to keep going. Designers should also pay attention to negative peaks and avoid reinforcing these as much as possible.

Last impressions are lasting impressions. In a cart funnel, if the most stressful steps are in the last part, the experience will be recorded as a negative one. To avoid negative final impressions:

- Split tough steps into substeps and give intermediate positive confirmations.
- Ensure harder steps are near the start of the journey, when users may be more engaged.

The goal indeed is to ensure each experience ends on a pleasant note.

In the end, the human mind is crafted to be efficient, which results in a loss of perfect perception and recall. We use mechanisms to optimize the storage of information by remembering the most intense points and last moments of an experience. So be careful about your ice cream quality, but don't forget to deliver a delicious cone as well!

Follow These Principles of Gestalt for Better UX Designs

Erin Malone

In 1910 a group of Austrian and German psychologists developed the principles of Gestalt psychology. *Gestalt* is defined by the *Oxford Dictionary of English* (Oxford University Press) as "an organized whole that is perceived as more than the sum of its parts." These collected principles (or laws) codify our innate visual perceptions and have long been taught to graphic designers and visual communicators as part of a basic design education. UX designers can leverage this perceptual understanding to create more meaningful, intuitive, and useful experiences by incorporating these simple visual principles into our designs.

The overarching *Law of Simplicity* states that people inherently group things together in their mind to simplify them and make meaning through that grouping. Our brains perceive patterns and complete shapes even when there is significant missing information. The following laws go into more granular detail about different ways that every UX designer can apply these principles to make designs more intuitive for users:

1. The *Law of Proximity* states that objects close to each other are easily grouped in the mind's eye. In UX design, if we group buttons and controls together, it's implied that these controls are related. For example, we can use this principle in navigation groupings and footer indexes to imply content relationships between items. We must be careful, though: if we group things that are not related, users will still assume a relationship because that is how we are wired.

2. The *Law of Similarity* indicates that people will automatically group controls and interfaces with similar shapes, color, and other visual qualities and apply common meaning to them—even if they are not physically grouped. Likewise with alignment: when elements are aligned, such as in

a left-aligned body of text, it is easier for the eye to scan down the page, whereas with centered text there is uneven and jagged white space on the left, where our eye starts reading, which creates visual friction, slows down reading, and makes the viewer work harder for the same information.

3. The *Law of Closure* is often seen in logos and icon design. People will perceive objects as complete even if there are gaps in the shape or line. This gives us a range of freedom when developing iconography for our applications and allows us to imply symbology or pictographs in more creative ways than straight representation. Classic examples include the IBM logo and the NBC peacock logo. A more contemporary example includes Material Design icons, specifically the icons indicating Off states.

4. The *Law of Common Fate* states that moving objects are perceived as moving along a path and that grouped items will be seen as moving along the same path. This is important when thinking about animating elements of the experience or using animation for feedback and within microinteractions. An example of this can be seen in a desktop OS when we select multiple files and then move them to another spot on the screen. The files appear to be moving along the same path as the cursor.

5. And finally, the *Law of Past Experience* states that people will use their past experiences to categorize and apply meaning to new experiences. We can take advantage of that perception by using common design patterns and templates in our work. People come to your experience with a host of past learnings. Leverage those past learnings for common tasks and interactions.

By leveraging what the human mind already does subconsciously, we can make our designs more intuitive for users. We can use the principles of Gestalt to our advantage, utilizing inherent perceptions to increase user understanding of our interactions and interfaces, allowing us to move our attention to solving more complex and nuanced problems.

Use Visual Design to Create an Eye Track

Kevin Lynn Brown

One of the biggest dilemmas users face when interacting with something is discerning where to start and where to go next. The visual design of an interface is pivotal to helping users answer these two questions, not only at an application level but also at a page or individual interaction level. UX designers can assist users through visual cues to identify key elements of the interface. But taking it a step further through the use of visual design techniques, it's possible to create an effective eye track or path for users to follow, thus leading them through the user experience.

At its base, design is a functional art, as opposed to fine art, which is an aesthetic art. There is overlap in that if a design is gracefully executed, it can have an artistic aesthetic quality as well—that is the pinnacle of design. This chapter focuses on the base level of design—making it functional and intuitive for users.

Whereas fine art communicates a message or an idea, design communicates a method or process. To communicate either an idea or a process by visual means alone, the designer/artist must control the viewer's eye. The designer/artist must decide what the viewer should look at and in what order and apply visual effects to achieve those goals. The base visual tools we have for this purpose are color, contrast, size, and shape.

When approaching a design, the designer first should decide where users should begin their experience and in what order task steps should be carried out. Then the designer must apply visual techniques to prioritize elements of the UI for the user's eye:

- *Text size:* Similar to how printed newspapers use headlines, subheadings, and bylines to prioritize reading information, text sizes can be used to sequence reading material.

- *Color:* In terms of color, warm colors advance forward, and cool colors recede toward the background, so simply warming up the color of initial steps and using cooler colors for later steps can create order. The degree of difference doesn't need to be extreme—just discernible.

- *Contrast:* Higher contrast indicates nearer objects and lower contrast indicates objects farther away. Simply increasing the contrast of initial elements and lowering the contrast of later steps can help lead users through a task. Contrast can be adjusted by raising or lowering the grayscale value of elements, or it can be achieved by increasing or decreasing the font weight of text without adjusting the grayscale value.

- *Line:* The human eye will follow a line until it terminates. If you bring a line in from the edge of the canvas and terminate it somewhere in the canvas, the human eye will always look at that point. Because of this, it's possible to use decorative lines in application UI to bring attention to headings, buttons, or other interactive elements that are part of the process.

- *Balance:* All of the aforementioned techniques ideally should be used in concert and used judiciously to have the intended effect. The user should not notice the variety of fonts or colors or be bombarded by lines pointing everywhere. Instead, the result should be that they intuitively are able to make their way through a UI task with as little friction as possible and without even noticing the visual effects in the UI.

Through the use of visual design techniques, it's possible to create an effective eye track for the user to follow, easily leading them through the user experience.

Use Object Mapping to Create Clear and Consistent Interfaces

Tim Heiler

Humans naturally understand the world through objects. According to Jean Piaget's work in cognitive development, most eight-month-olds have already discovered object permanence: they are aware that even if they can no longer see, hear, or otherwise sense objects, the objects still exist. The children have formed mental images of these objects.

Designers can leverage this knowledge not only to create more intuitive products by removing cognitive load but also to ease communication with their team and organization. Though UX designers like to think in terms of screens and user flows, thinking of the objects that make up interactions results in a level of clarity that your users, team, and stakeholders will appreciate.

Here are several observations about physical objects we want to remember:

- Objects in the real world don't typically change, but we might see them from different perspectives or in different conditions.
- Objects have parts and are themselves often part of other objects or categories.
- Objects in a category usually share similar properties.
- Certain parts and properties are more important to visual and mental recognition. For example, you will recognize a person by their face rather than by their fingerprint.

In your design process, you can leverage this knowledge through a series of steps:

1. *Identify the objects in your product:* Write down the objects in your experience that your user will interact with. For example, if a user story says, "As a user I want to join an event as a guest," the objects to identify would be "Event" and "Guest." Focus on the nouns used to describe the product you're working on.

2. *Assess the properties of your objects:* Add more information about how your objects are identified. The "Event" object will likely include data like "Time & Date," "Title," "Location," and maybe a "Preview Image."

3. *Assign the matching actions:* For our "Event" object, there will be actions such as "Attend," "Invite a Friend," and "Comment," among others.

4. *List the relationships:* In addition to the data already mentioned, the "Event" will also contain a variety of "Guests" that have registered to attend. Each "Guest" will be defined as their own linked object and have their own set of properties.

5. *Sort to priorities:* Once you have figured out the properties, actions, and relationships, determine the first, second, and third things users may want to know about an object. That list will likely resemble the result of a card sorting activity.

6. *Use your object map to build consistent interfaces:* Every time a type of object appears in your interface, whether it's in a list, on a card, or in a notification, use the prioritized information to display instances of the same object type in a consistent way. Also use consistent call-to-action elements according to your global priorities.

In real-life interfaces, users are often confused by identical objects depicted and labeled in inconsistent ways. This happens across screens, touchpoints, teams, and products, and throughout entire companies. Mapping, understanding, and communicating about these objects is an essential practice for every UX designer. Adopting this object-oriented approach will make your interfaces clearer, layouts more consistent, team discussions more focused, and various touchpoints more aligned to one another.

References

- Ansgar A. Plassmann and Günter Schmitt, "Das Entwicklungsstufenmodell nach Piaget" (The Piaget development stage model), Psychologie online lernen (Learn Psychology Online), n.d. https://oreil.ly/mxFEy.
- Sophia V. Prater, "Object-Oriented UX," A List Apart, October 20, 2015. https://oreil.ly/hLPzs.
- ———, "OOUX: A Foundation for Interaction Design," A List Apart, April 19, 2016. https://oreil.ly/4wJMy

Remember the Four Questions of Critique

Adam Connor

Critiques are a form of analysis. They provide participants with an understanding of how and why a solution might or might not meet its objectives. To accomplish this, critiques should do all of the following:

- Identify a *specific aspect* of the idea or decision under analysis.
- Relate that aspect to an objective or a best practice.
- Describe how and why the aspect supports or does not support the objective or best practice.

Four questions help ensure we include these details when giving or receiving critique.

1. *What are the objectives for the solution?* To analyze a solution, we need to understand the objectives it is intended to achieve and focus our examination in those contexts. Objectives include:

   - The audiences the solution serves (their behaviors, perspectives, needs, and so on).
   - The problems or opportunities the solution should address and the contexts in which they occur (when, where, and why).
   - The business and user goals or outcomes we want to achieve by addressing the problem or opportunity.
   - The beliefs, perspectives, and principles we want the solution to adhere to (What are the characteristics and qualities of the solution itself?). These can be principles the team has set as well as the best practices and tenets for design in solution space.

   In a critique, it's important to be explicit about which objectives are being discussed. Trying to analyze a solution against all the problems it should address, across all audiences, is often too much. When we focus on specific audiences, outcomes, scenarios, and so on, our conversations are clearer and more effective.

2. *Which aspects of the solution are we analyzing in relation to a specific objective?* Critique should specify the aspects of the design that are being analyzed in the context of the objectives. For example, we might be focused on a particular interaction flow, part of an IA, a screen layout, the visual presentation of a component or content.

3. *Is the aspect effective in working toward the objective or not?* The previous two questions provide a foundation for critique. Think of them as the elements on the left-hand side of the "=" in an equation. Now that we have them, we can ask: do we believe the design will work to achieve the specified objectives? Yes or no?

4. *Why is or isn't the solution effective?* And finally, it isn't enough to just answer yes or no; we have to explain why. Sharing this furthers the discussion and allows others to share their perspectives, whether they're similar or different. Gathering this range of perspectives helps us understand where we can iterate on a solution to explore improvements or build on strengths.

In addition to us thinking through these questions ourselves, they also help facilitate critique discussions with others. As team members share perspectives, these questions can help clarify and refine the points made. For example:

- When someone shares that they like or don't like a particular aspect of the solution...we can ask them to relate it to an objective (audience, scenario, goal, and so on) they feel it supports or conflicts with.

- When someone proposes an alternative idea for the solution...we can ask them to describe the aspect of the current solution they think isn't working toward its objective.

Keeping these questions in mind when we critique on our own or with a group helps ensure productive critiques. But while the questions may seem obvious, staying focused on them can be tricky and takes practice. The more intentional we are about asking these questions of ourselves and others when gathering feedback, the more useful our conversations will be in helping us iterate and improve the solutions we create.

Turn Poorly Constructed Criticism into Actionable Feedback

Jesse Nichols

As UX professionals, our job is to facilitate empathy and bring stakeholders together at the table. For this reason, it is important to solicit feedback at every step of the design process. But unlike our UX colleagues, stakeholders aren't always trained in the importance of empathy with others. Their feedback may come with unnecessary and maybe even insulting baggage attached. It's easy to take this type of feedback personally and disregard it. But when we do, we miss opportunities to improve our design. For this reason, designers must learn to parse out constructive criticism from even the most harshly delivered feedback.

The Big Bad Stakeholder

I've presented design work to countless stakeholders. Most of the time, their feedback is useful. But on occasion, I will receive a comment that feels more like an insult.

Is that because they are mean? Maybe. But more likely they are struggling to articulate around a complex topic. Unable to express their intent, they simplify their negative feelings into an unhelpful, generalized critique. Luckily, our experience with qualitative research has prepared us for this situation. In fact, we are uniquely armed with the skills to translate their feelings into actionable feedback.

A Simple Misunderstanding

I was once tasked with redesigning the home page of a website. When I presented the new home page style, I received the following response:

"I hate it. We should go back to the drawing board."

A younger version of myself would have done exactly that. I would have internalized my ire and retreated to my desk. I'd have discarded my previous work and started over.

But in this instance, I chose to dig deeper. Despite my wounded pride, I made the choice to assume positive intent. I set aside my ego and sought clarification.

"Let's talk about it," I implored, with an expression designed to communicate genuine curiosity. "Tell me more about what you dislike on this page."

As it turned out, they didn't dislike the design at all—they were fixated on the content, which was still in the draft stage! Had I given into my hurt feelings, I'd have thrown away a perfectly good concept and spent hours redesigning this page! Instead, I assured them we would address the content, and they approved my design.

Tips for Success

Parsing useful information from harsh feedback can be challenging. But there are a few things you can do to set yourself up for success:

- *Pause and take a few deep breaths*: Don't allow emotion to control your next actions. It takes only one passion-driven misstep to derail a conversation.

- *Assume positive intent*: Align yourself with the stakeholder. Remember that you are all on the same team. In most cases, feedback is not intended to insult you. The stakeholder has a concern, and they are voicing their opinion. They want to build a good product as much as you do! This is the time to set our egos aside and listen.

- *Ask open-ended questions*: This is not the time to walk away from the conversation. Instead, we want to open things up with questions like these: "Why is that important to you? Can you expand on your thoughts? Tell me more about that, please."

This is your chance to help them dig deeper so you can understand how to make your design align with their goals and values.

Feedback can be difficult to receive. It can be nonsensical at times and is often poorly constructed. However, it is imperative that we are able to uncover useful nuggets of truth among the criticisms we receive. After all, that's how we improve our designs!

Improve Communication and Encourage Collaboration Using Sketches

Anna Iurchenko

When synthesizing research findings and making sense of workshop outcomes, UX designers and researchers have to deal with an ocean of information and then effectively communicate insights to teams and stakeholders. And by communicate I don't mean that we write huge reports or create decks with diagrams and personas, though those tasks are still part of our job. Instead, I mean we have to facilitate understanding among the team and invite collaboration so others can build on our ideas, generate new hypotheses, and critique our decisions.

An ability to use simple hand-drawn visuals can allow you to reach this goal. Sketches help you bring others into your thinking process and make your ideas easier to grasp while allowing you to communicate to everyone in an accessible way.

What can freehand sketches do for you?

- *Simplify and strengthen your message*: Sketching naturally limits the amount of information you can put on the page, forcing you to cut everything that's not essential. Simplifying the narrative toward a purpose is an effective way to make it more memorable.

- *Make your ideas stand out*: We're all familiar with generic diagrams in which icons represent users and workflows are shown with blocks and arrows. By contrast, sketched aesthetics of your visualizations bring ideas to life in a casual yet informative way. They help you tell stories in a more loose and authentic manner.

- *Invite collaboration and co-creation:* Sketching by hand is seen as work in progress. That is part of its special power, and why it's so good at getting people talking. People see a rough sketch and instantly feel comfortable enough to compliment or critique it.

Here's how to get started:

- *Sketching is about courage, not talent:* Everyone was once five years old, an age when most of us were happy to draw any time. But then we learned how to express our thoughts and ideas using written words and abandoned visual language. Similar to writing, it is daunting to sketch at first; the only way to get comfortable with it is by practicing. Start simply by adding icons to your text notes—draw a star next to an important item, a bulb near the idea, or an eye for anything that needs reviewing.

- *Remember—good enough is perfect:* A sketch's primary function is to convey ideas just enough for others to build on them, so don't try to make your drawings too perfect. If people see your sketches as finished artwork, they'll be far less likely to offer constructive feedback, and you'll rob your drawings of their power as a co-creation tool. Keep your sketches about communication, not artistic expression.

- *Bring a sketchbook and pens to your next meeting:* Produce your notes using both words and simple graphics, but avoid the usual straight lines of text. Make sure that others see what you're doing. Eventually, your colleagues will start referencing your sketches, expanding on them, and building conversations around them.

- *Start by using skills you already have:* We live in an increasingly symbol-oriented culture, and lots of popular visual representations are simple enough to sketch; just think of all the emojis you see every day. Focus on making the intangible visual, and don't overcomplicate things. You just want to communicate your ideas as quickly and accessibly as possible.

Sketching is a very effective way to share and explain design decisions and encourage discussions. If you want to get your point across in a way that resonates with others, quickly engages stakeholders and builds energy, and encourages cross-functional co-creation, you should bring sketching into your design process.

See Part III, *If You Show Something Shiny, They'll Assume It's Done*, page 90 for information about presenting sketches to stakeholders.

Learn the Difference Between UX and UI from a Bicycle

Joe Wilson

It's essential to understand the connection between UX and UI before designing or developing products. One of the more memorable pieces of wisdom a brilliant creative director shared with our team explained the difference between UX and UI by using a bicycle. Here's how he broke it down.

UI: User Interface

The user interface would include things like these:

- *The seat*: Was it comfortable? Was it set to the right angle?
- *The handle grips*: Were they irritating? Too thick?
- *The brakes*: Were they too hard to squeeze? Too loose?
- *The pedals*: Were your feet able to reach them? Did your feet slip off of them?
- *The gears*: Was it easy to shift gears with your thumb? Did the chain get stuck?

The idea illustrated here shows how a user would interact with the bike's components to ride it. The bike example can easily translate to digital design components, which lend themselves to the same questions:

- Was the main call to action easy to find on the page? Was it large enough and easy to click?
- Was the font large enough and easy to read?

- Was there enough color contrast between the background color and the text color?
- Did everything fit on the screen, regardless of the device?

UX: User Experience

The user experience was the bicycle ride itself:

- Did the rider enjoy themself while riding the bike?
- Did the brakes work?
- Were they able to shift gears quickly?
- Did they get to their destination without falling off the bike or getting a flat tire?

If so, they had a good user experience while riding the bike.

When translating UX to digital design, we can ask similar questions:

- Did the website load quickly?
- Does the information architecture make sense?
- Is the the website or application easy to navigate?
- Were the design elements intuitive and easy to interact with?
- Did the customer journey ultimately take them to the desired goal?

If so, they had a good user experience while interacting with your website, application, or product.

It's helpful to think about this story every time we approach a new product launch, and about how all interaction elements add to the overall user experience.

We have the honor of working with exceptionally smart people in the design field who are always learning. Our agency's creative director was the same way. He was a brilliant man on a never-ending quest to learn and was addicted to input. As Joe Strummer said, "No input, no output." Our creative director lived by this ethos, and as smart as he was, he still felt like he did not know enough about the difference between UX and UI even after years of studying design. So he pieced together the story about the bicycle and then shared it with his team to make sure we were all designing products for the right reasons, which he made seem as easy as riding a bike.

A strong connection between UX and UI creates memorable customer experiences. Look at the digital transformation Domino's went through with its patented Pizza Tracker platform to enhance the customer experience. Not only can the customer choose the most convenient way to order (via Alexa voice, text, tweet, and more), but it's also effortless to order a pizza through the Domino's interface and track where it is on a map during its journey to your front door. What's more, while some may think Domino's pizza isn't anything to write home about, the company's ordering experience certainly is.

Exceptional user experiences start with effortless interface interactions. Designers who keep this in mind will always build great products.

Sell Your Design Ideas with Trust and Insights

Benson Chan

As you move through the design process, chances are you're trying to convince either yourself, the end user, or internal stakeholders (or all three) that your design solution is the right one. Delivering a product is a team sport, so the ability to sell your ideas is critical.

It doesn't matter if you're fresh out of a famous design school or you are a senior executive—no one is 100% successful with their ideas. I have seen many design reviews into which great ideas have gone to die and have also seen very bad design ideas somehow get shipped because of the quality of the pitch.

Frequent pitch mistakes I've seen include designers not clearly expressing what problem they're solving, not understanding how customers feel about their design, or being unable to predict internal stakeholders' concerns.

Here are four things to master to help you sell your design:

1. *Communicate your understanding of the problem and goal:* This sounds simple, but it often doesn't happen. Recap with stakeholders your understanding of the problem—rephrasing someone's concerns back to them in your own words is a common technique used to ensure you have a shared understanding. This also helps to ensure your stakeholders have confidence you're on the same page.

   To communicate your understanding of the problem and goal, start your design pitch with a slide called Design Goals, with brief but specific bullet points like "Reduce registration friction and improve our 40% drop-off rate." You can then detail the approaches you took to reduce friction in the design they're about to see.

   A method for rephrasing stakeholder concerns is to start with "What I'm hearing is..."; then make sure to address your understanding of why they're concerned and what you think they're saying you should do. For example: "What I'm hearing is that you think the drop-off rate is caused by one registration field and we should rethink that first before any major updates—is this correct?"

2. *Back up your design ideas with customer validation:* It's a best practice for designers to seek insights and data through primary or secondary research to help inform and improve designs.

   Along with your design work, plan to gather and communicate supporting insights, data, and best practice examples at every design review and explicitly call out the design changes you've made as a result of each insight. For usability insights, if you don't have a research team, you can use online UX testing tools to quickly run studies or conduct quick sessions with coworkers. See Part V, *Don't Underestimate the Power of Coworkers as Usability Participants*, page 189 for more information about conducting research with coworkers.

3. *Build trust with your stakeholders:* Pay attention to stakeholders' goals and general areas of concern. If you have a great understanding of your stakeholders, you can cater your research or design highlights to get their buy-in. You'll need to adjust what you pay attention to with stakeholders across disciplines like program management, engineering, design, marketing, and so on. See Part III, *Learn to Think like a Missionary, Not a Mercenary*, page 85 to read more about building stakeholder relationships.

4. *Show that you have both customer and business focus:* Successfully selling your ideas and having design influence also means having a strong understanding of the business strategy while keeping the customer in focus. For example, make sure your stated goal or goals are tied to customer and/or business outcomes. Then, for each stated goal, find supporting customer-centric and business-centric data points.

Fundamentally, it comes down to having a strategic and informed design process and how you communicate and bring stakeholders along. Use insights and data to inform your solution, and be deliberate in communicating how they support your design decisions. This will go a long way toward building the trust needed to sell your design ideas effectively.

Align Your Team Around Customer Needs via Design Workshops

Shipra Kayan

Early in my career as a UX designer, I tended to feel underrepresented in strategic decision-making. Everyone loved my design ideas, but they always fell into the "nice to have" bucket and were never prioritized.

The problem was not my ability to be convincing. The problem, I've since realized, was that I was fundamentally misaligned with my product and engineering peers about who our customers were, what problems we were solving, and the priorities of our business. I needed a new approach.

We had a big project coming up to redesign our sign-up and onboarding flow:

- Instead of waiting for a road map, I ran a hypothesis workshop with product managers, engineers, marketers, and customer success representatives. *Who is our target user? What is the user's journey? What are their goals? What are some stumbling blocks? What gives them pause?* We drew the user journey together and realized we had conflicting hypotheses about why our customers didn't convert.

- Instead of presenting findings from my customer interviews, I invited my team to a synthesis workshop. *What did you hear in that interview? What surprised you? What does this say about our assumptions? How should this new knowledge change our priorities?* We debated our inferences and came away aligned on the problem space.

- Instead of wowing everyone with gorgeous mock-ups, I invited folks to a sketching workshop so I could curate the best ideas. *Who does a good job solving analogous problems? Where can we find inspiration? How might we make this new process feel familiar to the user? What would you say to the user if you were the UI?* We were a tightly aligned team solving for the same user scenarios, and our work had business impact.

Once I started facilitating workshops, coworkers expressed relief about having structure for conversation so they could focus on the content. Soon thereafter, coworkers started asking me to facilitate road-mapping workshops and executive offsites, bringing the customer-centered design process into those conversations.

Here are three things you can do to start practicing design facilitation:

1. *Mindfully design every meeting you lead:*
   - Instead of giving a prototype demo and asking for verbal feedback, give people time to quietly absorb, write their thoughts, and share with others. Only after each individual has seen the design through multiple lenses can you have a rich discussion about conflicting feedback.
   - Instead of giving a research findings presentation, let your team explore your journey maps on their own time. Spend your meeting time having meaningful discussions about what surprised them or what they might do differently based on what they learned.

2. *Run a design sprint, ideation, or journey mapping session:* Try not to be "the designer." Instead, focus on creating space for other people to have insights, ideate, and design.

3. *Attend workshops led by expert facilitators:* Do this as part of a training or even with your colleagues who do it well. Observe their pacing, why they interrupt, and how they resolve interpersonal conflict. You learn the most by seeing others do this in a messy real-world setting.

Design workshops bring coworkers from different departments together to have meaningful discussions about customer needs and solutions. Don't assume you have to figure it all out yourself; facilitate a design workshop and harness the creativity of your entire team.

For more information about workshops, see Part III, *Make Learning a Part of Your Design Process*, page 94 as well as *Design Meaningful International UX*, page 96.

Embrace a Shared Cadence to Avoid Silos

Christy Ennis-Kloote

Complex product and service projects require *internal silos to be broken*. When everyone is focused on their own piece of the puzzle, they lose the ability to see how all the pieces fit together. Disparate teams lack intimacy gained from shared experiences, shared vision, or trust to fall back on. So when tough times do arise, teams don't have a *practiced cadence* or shared instinctual rhythm for effective communication—making difficult problems more difficult.

One multiplatform product my team worked on had five different partners at different companies, all initiated as separate requests from a prestigious customer. The partners worked as separate delivery teams with different cadences focusing on UX and UI design, mobile development, cloud development, and firmware development. Waterfall-esque requirements from the customer, provided to separate teams, made for slow builds and integrations. There was frustration all around at the pace and ability to see progress. This meant teams had assumptions and misses on what the other teams were doing. We believed the customer came with the same maturity in developing digital products as in their historically well-crafted physical products. It turns out, however, that this was their first launch in this space, and even with very competent partners, it was a challenge. While we played well with others, we respected the imposed boundaries—to everyone's detriment. We could have saved undue stress by applying baseline practices for healthy habits of communication. These practices increase cross-collaboration and are unique to the team composition:

- *Keep it brief:* First, start simple, with minimally disruptive requests for time from someone representing every aspect of the work, and don't seek perfection. Set dependable time together, and share what you are working on and any barriers to achieving those goals, clarifying any assumptions. Typically, teams do this through a call-in format called a stand-up, but even a text post shared regularly in a messaging platform for everyone on the team to see can work well.

- *Make it real:* For continuous delivery cycles, commit to incremental, demonstrable outcomes even if not the final deliverable (typically every two weeks, though some can be longer). For example, stakeholders looking for evidence could see how changes in a mobile experience engage customers. That data can be aggregated to look for information about where to pivot next. These pieces of evidence can help everyone identify dependency gaps and feel progress toward an objective.

- *Reflect:* Finally, for long-term goals and larger releases, include regular points to pause and reflect on the orientation of the team in reference to the goal so there is time to course correct. Each of these is a longer meeting to plan the work for the next two to four sprints. These meetings need to be held in an open format, with space to invite discussion around the larger dependencies that might impede the work.

Setting a shared cadence promotes regular contact across cross-functional teams. You will get to know each other by building a secure space to encourage transparency, as happened for us. It creates tighter feedback loops, keeping everyone current and oriented to the same near goals as a team. Regular meetings and habits will bring incremental change that can shift the direction over time toward achieving a truly impactful product before it is too late.

You can't influence impact on customer experience if you are not there to advocate for customers. So whether you are an outsider just coming through short-term or someone who's been with the product since inception, you have agency to choose how you participate. Embrace and invest in the whole team. They'll help you see what you can't to achieve better design outcomes together.

Learn to Think like a Missionary, Not a Mercenary

Scot Briscoe

Becoming a UXer is a life calling; we have schooling and practices, yes, but it feels like much more than that. It feels to me like a sacred order.

As UX designers, we accept the responsibility (and burden) of unveiling the mystery of UX to everyone in our organizations. Because UX is so misunderstood, it's tempting to see ourselves as mercenaries, ruthlessly cutting down anyone who we think stands in our way. Instead, our role is to be missionaries—evangelists of UX—to everyone around us. Spreading the gospel of UX is about advocating for users and advancing the cause of UX in the process.

If you work in UX, you know that our craft is equal parts science, art, behavior, and data. As a practitioner in this innovative field, you must do more than just applying tools to user stories; you also need to carry the torch of our practice.

Pitfalls of Mercenary Work

A mercenary approach to UX is not collaborative: it is often motivated by a desire to "win" and a sense of being right, often accentuated by short-term disagreements. By contrast, missionaries know the importance of short-term gains but are fundamentally driven by long-term strategic goals. To achieve those goals, missionaries strive to achieve small wins *toward* that future ideal state.

To that end, don't let your idealism for doing right by the user prevent you from getting heard. Instead, learn to use the language of your business to advocate for your customers.

When your CEO pushes back on a design or complains about the time it takes to do research, resist the urge to feel defensive. Instead, acknowledge it as a different vocabulary. This may be the only way they have to express a business need. Instead of preparing for battle, find the opportunity for a dialogue; instead of trying to be right, consider their perspective and frame your request in language and data that resonates with them.

Becoming a Missionary

Acting as a UX missionary means thinking strategically at all times while building your narrative and support for that vision. Here are some ways to progress toward becoming the UX missionary in your company:

1. *Build respect for UX in your organization:* Practice *design doing*; create value metrics for design, share studies and findings across your organization, and involve stakeholders and leaders in an ongoing engagement model.

2. *Advocate for the customer:* UX is not the only voice in discussions about solutions, but we do have a duty to put customers first. Since not every role has that same customer view, we must always advocate for them.

3. *Find your followers:* There will be natural points of alignment and partnership with tech and product, but don't forget to build relationships with folks on other teams. Reaching across your organizational aisles will help you build trust and find partners.

4. *Solicit advice:* Create your own board of directors by actively cultivating mentors from among colleagues, peers, and contacts. Ask for input on your thinking or for help on specific topics. I have multiple mentors, each focused on a specific area like design, leadership, technology, and so on.

Stay on the Path

During my own journey along this winding UX path, I've failed a lot. Expect to stumble on your missionary path, and to wander off the path once or twice. You won't nail it every time, but a few short-term losses don't negate the bigger picture.

As a UX missionary, I wake up each day excited to share my craft with the world. Seeing others get the UX spark also makes it worthwhile; seeing that spark spread throughout a company is amazing.

For more information about selling design ideas, see Part III, *Sell Your Design Ideas with Trust and Insights*, page 79.

Not All Interfaces Need to Be Simplified

Morgane Peng

Whereas consumer products are aimed at the individual, enterprise products cater to complex organizations. They are designed for business professionals or employees.

A common temptation is to oversimplify these enterprise products—I see this mistake in many design interviews.

Get Familiar with Enterprise Products

To understand what "good design" is in the enterprise space, you have to get familiar with enterprise products and understand the difference between business and interface expertises. This will help you find the sweet spot between the "not too simple" and the "not too complicated."

Every mainstream app has an advanced equivalent for its professional users or internal employees: travel agents, visual effect artists, financial traders, and so on. Their applications are tailored to their specific professional usages and are complex for a reason. These users often need to see a lot of information at a glance to quickly make informed decisions, compare data points, or see the status of multiple systems at once.

When we designers try to oversimplify an interface, we are not enabling efficiency for the users: we are actually making the product more difficult to use by obscuring interactions and data!

Differentiate Business Expertise and Interface Expertise

The key lies in understanding the difference between business expertise and interface expertise:

- *Business expertise:* A person's knowledge in an area or a topic due to their studies, business training, or work experience. It also includes the shared mental models on how they approach their tasks and processes.

- *Interface expertise:* A person's knowledge of how an interface works due to their training, habits of using it, or experience from using similar interfaces. As their knowledge increases, so too does their ability to recognize familiar interaction patterns instead of having to analyze or recall from memory.

Find the Sweet Spot

Let's look at the three points in the graph shown here:

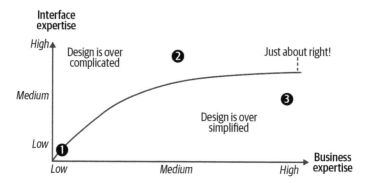

1 This is where products with a large consumer base—apps for ordering food, listening to music, booking a train ticket, and so on—aim to be.

2 This is where a lot of legacy tools are: designed with the system in mind, not the users. Projects assumed that business expertise meant interface expertise.

3 Danger zone! The interface is dumbed down too much. Users now interact with a black box.

The goal is to be somewhere along the curve. Over the years, I started to take a portfolio approach with our projects at work: with different target sweet spots, we give the right level of complexity to the right people.

To get started on finding your project's sweet spot:

- *Don't be lazy on user research:* Understanding people's context will help you choose relevant design patterns.
- *Design with people, not for them:* By opening the design process, you can coach stakeholders and study participants to help you break down any complexity.
- *Be gradual:* Adjust and introduce interface complexity gradually, with personalization and user settings, or with features like shortcuts and keyboard navigation. Users with lower business expertise can ignore those interface capabilities until they become more proficient.

As Don Norman said: "Life is complex and tools must match life." More than ever, we designers are needed to tame the complexity of life.

If You Show Something Shiny, They'll Assume It's Done

John Yesko

UX designers tend to follow something like this general path when documenting a digital user experience:

> user flow > sketch > wireframe > visual design comp > prototype

We may start with a rough flow or sketch to get alignment on the general concept. Then we design wireframes to add detail to the interactions, and we start suggesting the layout while documenting the non-"happy" path. Ideally, we also conduct user research along the way to make sure the experience is meeting customer needs. Once we feel good about all that, we create a polished visual design—either a static comp or an interactive prototype. Despite any nuances in our individual processes, the important theme is that the fidelity of the deliverable increases as the details of the experience take shape.

Sometimes, though, we fall into the trap of presenting a finished-*looking* design when we're still in the early stages of thinking it through. This often happens when we're trying to impress executives with "what could be." We assume that they will respond only to a high-gloss interface and that they can't be burdened with the less glamorous design work that must happen first.

Our stakeholders might walk away impressed, but this approach has pitfalls. For all the "work in progress" caveats and "FPO" stamps on screens, polished designs give the impression that more of the work is done than truly is. The "thinking" part of design may be 20% done, but the interface looks 80% done.

In these cases, a glossy presentation undermines the real nuts and bolts work of UX design—solving problems. When the design work appears to be more finished than it really is, our stakeholders might assume that we can jump right into development and quickly launch the product. Although there are likely still many problems to solve, we've created the impression that our design is ready to be brought to life.

So how do we get around this landmine?

- If at all possible, try not to show polished design work too early in the timeline. Sometimes the work process or culture won't allow that, but give it a shot.
- Orient stakeholders to the broader design process by illustrating the high-level steps, along with providing a clear "you are here" indicator.
- Let the audience know what *kind* of feedback you're seeking at this stage, appropriate to the current level of fidelity. For example, head off discussions about specific text if the copywriting hasn't been finished, or about color if you haven't thoroughly explored that area yet.

Keeping in mind that stakeholder management can be just as important as hands-on-keyboards design, these steps can help set realistic expectations for the remaining work. Not only do we hope to cement the UX process, but we also might head off tough interactions down the road. The big win is raising the profile and understanding of UX design by elevating discussions beyond pretty pictures.

For more information about sketching, see Part III, *Improve Communication and Encourage Collaboration Using Sketches*, page 74.

You Can't Always Help Who You Want

James McElroy

Helping your coworkers deliver better customer experiences can be bad for your reputation as a design leader.

Sounds crazy, doesn't it? This chapter explains when it's time to stop selling great UX and start focusing on delivering great UX. Spoiler alert: focus on delivering excellent experiences for the people funding you instead of delivering mediocre experiences for everyone.

Especially when design leaders work in an organization with a low level of design maturity, we tend to evangelize the value of good design by helping anyone who will listen to us. This often means taking on side projects, volunteering our time, and working in areas outside of our primary team.

By helping solve other teams' problems, we build credibility and increase demand for our services. To be clear, this process typically takes years, and the results are uneven; some teams increase their design maturity far faster than others.

If we're successful (and perhaps a bit lucky), business leaders see the ROI of good design and begin shifting some funding tied to solving specific business problems to design so that we can help solve those problems.

In the early days of UX at many organizations, that help can be easy to deliver—if you're starting with a terrible customer experience, the fixes are obvious.

But once you fix the obvious issues, things get harder. This is the critical inflection point many leaders miss (I know I have). If your designers are still spending their time (which is now your sponsor's money) trying to sell other people on UX, they're not focusing on delivering great experiences to your most important customer—the sponsor who has already been sold.

And here's another challenge: designers tend to want to help people, especially people who are trying to solve interesting and important customer problems. It's hard for most of us to turn down requests to solve those problems, and it's even harder to watch a company release poor customer experiences your team could improve.

But if those problems aren't the ones your leaders have prioritized, your reputation depends on letting those poor experiences go out the door.

Here's why: as soon as your design team reviews and provides input on a design, your team's reputation is associated with that design. Even if you just "took a look" at mock-ups for a product that needs a fundamental redesign, well-meaning product owners will market the fact that the UX team "reviewed" their design, implying the design team approves of it.

At this point, you're stuck. Making the product good requires a heroic investment of your team's time, sacrificing the product you've been funded to work on (and likely any semblance of work-life balance). Staying focused on your product results in your team being associated with a bad product.

How do you prevent this?

- Focus your team on priority work, as determined by your sponsor.
- If the team coming to you believes their work has a higher priority than the work your team is focused on, have that discussion with your leadership and ask them to choose.
- Let the difference between the projects your team works on and those your team doesn't work on sell your team. Bad product releases can actually make the design team look good.

It's hard to watch bad designs go out, but successful design teams always have more demand than they can meet. Don't damage your hard-earned reputation by overextending your team; delivering great results on your priorities is one of the best ways to get the funding and staff needed to take on that other interesting work.

Make Learning a Part of Your Design Process

Michelle Morgan

Most people focus on learning from mistakes, but the bulk of our daily work comprises things we consistently do right, things that are neither right nor wrong, things not yet done, and new things that resulted in successful experiences. As UX practitioners, if we take time to reflect across all aspects of a project—the good, the bad, and the meh—we can exponentially increase our learning and self-improvement. Here are some ways to incorporate learning into your work:

1. *Keep a daily work journal*: Jotting down a few notes at the end of each day creates a habit of review and assessment with low effort. Use a specific notebook that nothing else goes in and keep the questions to a standard sprint framework:

   - What happened today—tasks by project, unusual events, insights?
   - What should I stop doing? Continue doing? Start doing?
   - What do I need to learn?

2. *Analyze a recently completed project*: Before a sprint or project retrospective, take 20–30 minutes to review and identify significant events. Start with your calendar to establish a chronology. After noting what happened and when, ask yourself a few questions:

   - What was great or fell flat?
   - How did the client react?
   - What would I do differently?
   - How is similar work in the office going?

3. *Review a coworker's projects*: Have someone review their project with you. Take time to digest the drawings and notes. Ask questions and imagine how you would have done the work. Add new ideas and methods to your notebook for reference. Make sure to note areas in

which the coworker is especially skilled for later advice. To take it to the next level, think about what happens when you switch teams and projects, or consider other roles. For this, I refer to the UX Checklist (*https://oreil.ly/EBkEk*), which is readily accessible on GitHub.

4. *Look for patterns across a few projects, especially the very good and the very bad, and ask questions like these:*

   - Are there places the team slows down?
   - How does the client respond to information?
   - Does one team always ask the same questions during handoffs?
   - How can I get better input in reviews?

     Then shift to solutions and capture improvements to make. This might include critical path contingencies, checklists for specific steps, or a mental model that helped develop solutions. I always use a range of sizes (S, M, L) when scoping a project to force thinking about what can be trimmed or added to the process. It's important to give some thought to team dynamics and to what is happening in the background—in your life, in the office, and in the world. Team members bring specific skills, and stressful external events affect the effort. Sometimes stress applies pressure that results in greatness, and sometimes it's unnecessary static. Making notes helps you recall issues and possibly avoid them next time, or it points to exceptional contributions by teammates.

5. *Create a learning process for the organization:* After identifying patterns, it's important to determine what needs improvement and what can be leveraged as a unique value of the team. Asking questions from management's viewpoint may identify new opportunities:

   - How does this work fit into our business goals?
   - Are we doing our best work for the best clients?
   - Where are our team's gaps? What are its unused skills?
   - Are we getting better?

Designers transfer solutions across projects every day and should look at this as a transferable skill to be applied as a learning strategy. Doing so improves team maturity, client experience, handoffs between teams, and project communication. Most importantly, it normalizes a culture of learning. It's challenging, but discussing concepts, strategizing improvement, and planning a different approach helps everyone deliver better work over time.

Design Meaningful International UX

Yingdi Qi

Our products are no longer used in single markets. Internet accessibility and digital innovations have reshaped the global user landscape. Nearly 60% of the world's population is online (*https://oreil.ly/Gw85C*), and 74% of those people do not speak English (*https://oreil.ly/54RSe*).

To drive user retention and business impact outside of your home market, it's crucial to craft a user experience that resonates with an international audience. *Resonance* goes beyond "translating the UX copy." It requires a deep understanding of the local users' behaviors and their contexts of use to inform design efforts.

A while back, my team undertook the design of an identity verification flow. By talking with local users, we realized we had faulty assumptions: although identification cards are pretty common in the US, a large number of our worldwide audiences do not have a valid ID card. That wake-up call showed us that if our solution were to focus on uploading a photo of one's ID card, we would exclude millions of users from registering. Engaging with international users up front helped us explore more accessible options early on and saved us 10x the cost of having to shift gears later in the cycle, let alone sparing us the pain of missing out on millions of global users.

Designing for international users is a fun, humbling, and rewarding challenge. Here are three approaches to set you up for success:

1. *Design with local users in mind:* Empathize with users via international lenses by doing regional/local research in parallel with the research in your home market. Gather input from local users about the on-the-ground realities. Understand their behaviors, values, expectations, and context of use. What perfectly suits one culture may be quite offensive for another. Doing local research at an early stage helps validate

hypotheses, gain fresh perspectives, and minimize failed design outcomes (and brand damage) from the get-go.

2. *Engage your localization team early on:* Localization refers to the adaptation of a product to fit the needs of the local audience. Successful localization requires an alignment of context. Without context, even the most experienced linguists may misinterpret things; consequently, local users may get something that totally deviates from your intention. Engage your localization partner or local experts early on, or even better, involve them in your design process. Empower them with context such as:

   • Business problem and goal

   • Persona

   • Target markets

   • User flows

   • Tone, voice, and style considerations (for example, unleashing creativity versus staying loyal to the original copy)

   • Anything learned from regional/local research

   Tap into their insights to further reveal blind spots. Seek their help in adapting your prototypes to target locales.

3. *Test with local users:* Test localized prototypes with potential local users of your product and conduct the research in their native tongue. Gather their feedback on localized user journeys. Validate whether everything from iconography, feature name, and language to the overall flow/solution appears "natural" enough, as if crafted by their favorite local companies. Incorporate learnings into your design.

Meaningful cross-border digital interactions rely on a design that looks, feels, and acts familiar to a local user's eyes. To drive up user engagement, a brand needs to fuse empathy for local users into design decisions and executions. Through the three approaches outlined here, you will be well on the way to embedding international thinking in the foundation of your UX process.

Legacy Product? Imagine You're Restoring an Old Farmhouse

Christopher Coy

The right attitude can make a huge difference when you've inherited an existing product. An analogy I use that helps positively frame the challenges of legacy systems and software is to imagine that you're restoring an old farmhouse.

Like the farmhouse, products need regular upgrades and improvements to keep pace with the world as it changes around them. Even a small repair, like adding a new electrical outlet, might reveal an entire rat's nest of old wiring...and turn a simple afternoon project into a weeklong undertaking. Tackling and patiently responding to these types of design and technical inheritances is hard—even for seasoned experts. As designers, we can leverage and anticipate these moments to model how UX is vital to improving the process and chance for success.

Here are seven strategies to address legacy product design:

1. *Embrace the mess from the outset:* You're typically dealing with a tangle of code, product decisions, and temporary solutions that unintentionally became permanent. Accepting these factors as part of the design problem you've agreed to tackle will help you not get so frustrated as you seek to unravel them.

2. *Bring key players together early and get them aligned:* If you can't get stakeholders to show up or care, it likely means you're focused on the wrong goal/problem...or the initiative is not important enough to the organization to have the support it will need to ultimately be successful. A design thinking strategy workshop with an external, neutral facilitator is a great activity that forces this agenda.

3. *Validate early assumptions through quick, preliminary user research:* Often you'll have to defend the value of this research as a tactical imperative worth the investment and time. Done right, it helps avoid knocking holes in the wrong walls.

4. *Use concepts and early design artifacts to get people excited and willing to go along with the project's vision:* Remember, it can be difficult for others to see potential in the early days. It's a major reason why architects make conceptual renderings and models. Don't forget to communicate major changes and challenges to these concepts broadly and loudly so that stakeholder expectations are managed appropriately along the way.

5. *Understand that some individuals will wait out the first phase on the sidelines, offering opinions without helping:* It serves them politically to do so. Accept that once the hard work is done and the project is looking successful, they'll turn up as part of the photo opportunity. This is OK—just be sure that you've cataloged the true work of those who believed, supported, and led from the very beginning.

6. *Pad your early time, money, and resource estimates:* Things seem to invariably go sideways at some point. It helps to have a built-in buffer.

7. *Anticipate that education and internal storytelling are a continual job, not just something that bookends your overall project:* The rest of the organization needs to hear about all the hard work and problem-solving you've been up to in order to value the price tag of time and effort.

Remember: living in the midst of a difficult restoration project is also part of what it means to inherit a legacy. The early choices of others, however backward they may seem to us in the present, represent their sacrifice and commitment to a vision of the future as they understood it. As stewards of the now, it's our responsibility to do our best to enhance that legacy and build something even better for future generations.

Be Your Own Project Manager

Tripta Kumari

When you start a new project, it's tempting to launch straight into the exciting stuff: the research, the ideas, the prototyping, and so on. But unless you take responsibility for planning the important elements and timing of your work up front, you risk losing control of your own processes. Ultimately, this can slow you down, stress you out, and make you lose the all-important space to flex your creativity.

I've spent many years of my UX career being parachuted into different projects, often with very little time to familiarize myself with preferred project methodologies. I used to struggle to get my head around questions like "How long do they expect me to spend on the research phase?" and "When will they expect to see user flows?"

Acting as your own project manager doesn't have to be painful or involve loads of detailed Gantt charts. By developing a templated project planning schedule, I've created an easy way to hit the ground running, regardless of the setup and constraints I find myself working with. It's a document you can refer to again and again during the course of a project. I even presented it at an interview as evidence of the way I approach a new project. I'll never forget the impressed look on the hiring managers' faces.

To create your own project plan, consider including these four key elements:

1. *A high-level project brief, with answers to questions like why we are doing this, who it is for, and who the key decision makers are:* It's surprising how easy it is to lose sight of these questions once you get into a project; it helps keep you on course when you have them where you can see them.

2. *A basic plan showing the main stages of the project (discovery, first draft, feedback, and so on):* This allows you to see at a glance what needs to be created, when it needs to be delivered, and how long stakeholders have

to give feedback. Once populated with dates and deadlines, you'll be able to timebox your work effectively.

3. *A reminder of the problems you need to solve, the information you need to gather, or the artifacts you're expected to deliver at each of those stages:* For example, your first concepts might need to show your initial ideas but also catch any misunderstandings. This helps ensure you meet the project team's expectations.

4. *A detailed plan that breaks down the work you'll be doing each week or even each day of the project (useful depending on the project's complexity):* For those projects where it's useful to be able to demonstrate what you've been working on, this can also act as an invaluable audit trail.

Having all this information gathered in one document is incredibly useful, and adapting your own custom template for different projects can save a lot of time. Be open and share it with stakeholders to help confirm their expectations early on.

Use the project plan as an ongoing reference point for what you're trying to achieve, a check when you're in danger of spending too much time on one stage, and reassurance for yourself and others that you know exactly what you need to deliver and by when. By taking responsibility for your own project management, you'll feel more in control and ensure you're able to leave enough time for the creative work you do best.

Design for Users, Not Usability Studies

Aaron Parker

Usability research is a fundamental part of user-centered design, but it can be a double-edged sword. An increase in usability research can correspond with a decrease in creative risk taking. Over time, we learn what patterns test well, and we start to rely on those patterns more in our design work, resulting in safe but potentially uninspired designs.

It's easy for this risk-taking aversion to sneak up on you. I once worked with a client whose business I'd gotten to know really well. I'd spoken to somewhere between 150 to 200 of their customers—running interviews, testing designs, and so on. It had gotten to the point where I probably knew their customers better than they did.

It was a few weeks into the project when our attention turned to the website's navigation. I put together a few different designs and was confident they would test well based on past usability studies. Another designer, however, decided to go with something much more creative. He'd designed a full-screen navigation system with a lot of room for images and copy. I wasn't keen on it, and I adamantly gave him many reasons why it wouldn't work. When we tested it, however, the participants loved it.

That's when I realized how safe my work had become. Instead of pushing boundaries, I was trying to fit within the established mold. I wasn't using my experiences creatively, and my thinking had become restricted, maybe even boring, as a result.

I've seen this happen a few times now with different designers. They get a few rounds of usability research under their belts and then their creativity starts to drop. Granted, it feels professionally validating when your work tests well and pretty disheartening when it doesn't. But if this becomes your focus, you quickly stop designing for your users and start designing for your test sessions.

Usability research is the safety net that allows you to be creative. It gives you the opportunity to experiment with new ideas, learn what works well, and catch what doesn't. But if you always stick to tried-and-tested design solutions just so you get acceptable results, you heavily reduce your opportunity to learn. Playing it safe also won't help you discover exciting new solutions that come only from daring to do something different.

So what are some simple things you can do to encourage more creative risk taking?

- *Embrace riskier design decisions as part of your design process:* Aim to come up with a mix of safe bets and long shots, and then bring at least one of each into your research.

- *Get a second opinion:* If you think you're playing it safe, ask a colleague to review your work. Get them to challenge the creativity of your design decisions and identify where you could experiment more.

- *Set expectations for your team and for stakeholders:* Don't let colleagues' pressure to stay low risk result in designs that always perform well during research. Instead, ensure your coworkers understand that part of the process involves taking design risks so that you may learn from your mistakes.

- *Don't think of usability studies as something you pass or fail:* The insight gained from a design that doesn't perform well can be as valuable as finding a design that does. It will help you make better design decisions and can grow your expertise as a designer.

The next time you're conducting a usability study, sit back and reflect on your designs. Have you tried something different and creative? Are you experimenting with something new? Or have you played it safe and avoided taking any risks?

Frame the Opportunity Before Brainstorming the Solution

Brian Sullivan

When you start any design project, identifying a specific opportunity is the single most important factor for success. If you are addressing the wrong opportunity, you are not giving your users what they need. Framing the opportunity before brainstorming solutions helps your team focus on what's important. It's the starting point for all future discussions.

Design should be intentional. It should solve a specific user issue or pursue a business opportunity. Too often, we see hidden agendas, promises already made, little (or no) user research done, contracts signed, or solutions predetermined. Framing the opportunity was an afterthought, not a forethought.

A few years ago, a product manager came to a design team wanting a dashboard. She told the team her customers could have all their data on one page. The product manager did not fully understand the problem. Luckily, the designers challenged the initial design challenge using a simple framing statement called WHO-WHAT-WHY. Here is an example:

> A (person) needs a way (to solve a unique problem) so they can (achieve this measurable outcome).

People resist change, but they want to improve. As a framing device, WHO-WHAT-WHY statements help you focus on what motivates a person to change their current behavior. *Why* leads the way for future brainstorming efforts.

Let's see how to use WHO-WHAT-WHY on a project—a hotel agent needing to see information about packages arriving at their hotel. We will start with a poorly framed opportunity and improve it using the WHO-WHAT-WHY format. Here is the first attempt:

> A (hotel agent) needs a way (to see a dashboard) so they can (see all their data on the same page).

The statement is poorly framed. First, there are different hotel agents: front-desk, reservations, back-office, catering, and so on. Second, the dashboard is defined as the problem (it's actually a possible solution). Finally, there is no measurable outcome.

Here is a second attempt at framing the opportunity using WHO-WHAT-WHY:

> A (front-desk agent) needs a way (to see packages arriving today) so they can (let their guests know right away).

The second attempt is better. The focus is on a specific person (a front-desk agent) and their unique problem (the need to see packages arriving today). The dashboard is no longer in the opportunity statement. But the outcome is not measurable. What's the impact we want to make?

Here is a third attempt using a WHO-WHAT-WHY statement:

> A (front-desk agent) needs a way (to see packages arriving today), so they can (let their guests know within 10 minutes of their arrival).

This attempt shows a measurable outcome (letting guests know within 10 minutes of arriving) to focus your team on why you are doing the project. Your team is focused on an outcome that should resonate with a specific person. Framing the opportunity is the starting point for all future discussions. *Why* leads the way!

Be Wrong on Purpose

Skyler Ray Taylor

What if design acumen were measured not by *how perfect your work is* but by *how willing you are to be wrong*? Whether it's with a client or within your team, an imperfect solution could be the catalyst that inspires your best ideas.

The Wrong Answer

I once spent a week with a client not making any progress on a cross-country delivery application. Our UX team needed to nail down the flow so we could begin wireframes, but the client was having trouble committing to what was *in* and what was *out* for the first generation of the product.

After days of pitching ever-expanding solutions and ending up in circular debates, I decided to attempt something new—instead of trying to be right only to get it wrong, what if I tried to be *wrong* in order to get it *right*? I intentionally started our next session by writing an incorrect assumption on the whiteboard. I wrote that the users of the app knew the exact time they would arrive at a given destination so they could send a manual alert for when the delivery was expected; in reality the system should calculate the arrival time based on their location and send automated notifications. The client and the design team pointed out the error and began to collaboratively make corrections. A healthy debate started, and soon the board was filled with arrows, circles, and dotted lines. By presenting something I knew was "incorrect," the perfectionism that had plagued the creative process was squashed. We created a clear set of requirements and moved to the next phase of design.

The Right Time to Be Wrong

Willing to be wrong on purpose is about removing your ego from the process and loosening everyone up so you can move forward. The next time you find yourself in one of the following situations, try stating an incorrect assumption or idea as a method for getting unstuck.

When You Feel Pressure to Get Everything Just Right

Perfectionism can stymie progress. By focusing on incorporating *all* the details into a design, you can become overwhelmed with inputs, unable to get them all into a cohesive whole. Propose a solution to a portion of the problem with the full knowledge it won't solve for everything. Small victories add up.

When You Don't Have Time to Be Wrong

Waiting until you have the perfect solution in mind usually leads to more waiting, followed by even more waiting. It takes only a few moments to be wrong. Create something. Anything! Bring an unsuccessful design to your next critique and have the courage to let your fellow designers rip it apart. You will very likely walk away with some ways to make it work.

When You Think You Are Always Right

Sometimes you get too comfortable in your own skin and need your assumptions challenged. Several times in my own career I've rehashed what worked last time, only to be frustrated that it doesn't work now. Intentionally choosing a different approach, even if you have a hunch it won't work, can get you to start looking at the problem in a different way and prevent you from becoming a victim of your own success.

Don't Stay Wrong for Long

Being wrong on purpose is not about *staying* wrong on purpose. It's about being brave enough to be vulnerable. May you become comfortable with not having to prove you are the smartest person in the room—you are so incredibly smart, after all!

Create a Lasting Design System

Lara Tacito

A design system can mean many things, but it usually consists of three artifacts: a style guide, a component library, and supporting documentation. Generally, a design system enables teams to efficiently create consistent, learnable, and cohesive experiences for their users.

If you're just getting started with design systems, there are many resources and tools to help create the needed artifacts. Though creating them isn't a small task, they have become easier than ever to produce. Instead, the main challenges are getting people to adopt them, keeping them up to date, and improving them as your team changes. Here are a few things to keep in mind as you create, maintain, and evolve your design system.

Make Your Design System Easy to Use

Treat it like any purpose-built product. Apply your UX research skills and gather information from users. Who is going to use it? What are they trying to do? How do they do this today? Where are the biggest pain points or opportunities to make it more efficient or remove friction? Make your design system self-service, not full-service:

- *Integrate into existing workflows:* It's hard to get people to change their behavior. How can you meet people where they are? What's intuitive to them already? Where do they go to find information or resources today? If you already have a design system, interview people who use it and map out their journeys. Find key areas where you can improve the workflow to encourage adoption.

- *Don't silo information:* Give designers visibility into how components work and how things get built. Help engineers understand how the components should work in context. For example, you could consolidate your design and engineering documentation to empower anyone to make informed decisions for the user.

- *Find ways to automate:* Are there rules you can build into components? How can you remove decision points but allow for creative flexibility? If the guidelines are built into the system by default, chances are people will follow the path of least resistance. For example, if there's always 10px between buttons, build that into the component. This avoids the designer specifying this each time, and the engineer doesn't need to guess.

Create a Process, Not a Project

Unfortunately, you can't just create a style guide, components, and documents and then expect to be done with your design system. Like any product or tool, it needs maintenance and improvement over time. Things need to be added, changed, and removed. In time, a process will emerge, designed or not. Apply your UX design skills and decide how the process should evolve as you create your artifacts:

- *Identify roles and responsibilities:* Clarify who identifies, approves, and implements changes for each artifact. Set and document clear expectations of those people in each stage of the process.

- *Create visibility into prioritization, decision making, and status:* People can lose trust in a system and will avoid using it if they don't know what's currently happening or how it works. Make the process as public as possible. Document decisions and the reasons behind them; this can serve as a reference over time or for new team members.

- *Communicate changes:* Find the best way to communicate updates to the team. This could be through email, an intranet, or whatever works best for your team. Ask for ideas and feedback on how they want to learn about changes. Make sure communication is clear, timely, and consistent.

Improving how an experience gets built impacts its quality. Don't let your design system become stale or ignored. Use these methods to help your design system remain a powerful tool in creating a great user experience.

Your First Idea Is Sometimes Your Worst Idea

Audrey Bryson

Imagine you're tasked with solving a design problem without an obvious solution. After examining the constraints and thinking deeply about the problem, an idea pops into your head. This idea seems reasonable from both a usability and a development perspective, so your work is done, right? Unfortunately, there are some serious drawbacks to running with your first flash of insight, and we should all learn to value the importance of multiple concepts.

There's a well-documented reason we, as humans, are so quick to go with the first viable option: our brains are lazy. Well, our brains aren't lazy, but rather we have evolved to conserve as many mental resources as possible. In his book *Don't Make Me Think* (New Riders), Steve Krug refers to this concept as *satisficing*: "Most of the time we don't choose the best option—we choose the first reasonable option." Other researchers have described this concept as the Einstellung effect. Bilalić, McLeod, and Gobet (2008) found that "the more familiar solution induced the Einstellung (set) effect even in experts, preventing them from finding the optimal solution."[1] By fixating on the first reasonable solution or the most familiar solution, designers not only miss out on flexing their own skills but also risk missing out on the best possible ideas.

I discovered an example of this principle recently while translating a desktop application to a tablet platform. My goal was to make many of the existing features available on tablets. During the project, our team experimented with the copy and paste feature—more specifically, we wanted to provide users

---

1 Merim Bilalić, Peter McLeod, and Fernand Gobet, "Inflexibility of Experts—Reality or Myth?," *Cognitive Psychology* 56, no. 2 (March 2008): 73–102.

the ability to copy individual images and paste them to either the same canvas or a different canvas. Most designers recognize copy and paste as a foundational feature of computing technology and one that does not need reinvention. We could easily have followed the familiar desktop pattern, but we felt that this was inconvenient for users who wanted to paste an image to multiple canvases, especially on a touch input device. Instead of following the original copy/paste workflow, we continued to iterate on ideas until we came to an alternative that we call the "copy to" feature. This idea allows users to easily send copies of their images to as many canvases as they want. Not only had we improved on the feature set for our product, but the product manager remarked that this would be an excellent addition to the original software. Without reconsidering our first idea, we would have missed the chance to enhance our project and software.

My advice is simple: don't stop brainstorming as soon as your first reasonable concept appears. Creativity is the product of looking at a problem from many different angles. Set an arbitrary number of ideas (3, 5, or 10 ideas can all make for a good starting point) to dream up until you start evaluating each concept's potential. Ask colleagues to provide feedback, as their input can point out obvious flaws or new possibilities. Sometimes you need to push past the first obvious idea to uncover the best possible solution to your problem. Of course, your first idea might not be your worst option, but it might not be your best one either.

Question Your Intuition and Design to Extremes

Navin Iyengar

There's a persistent myth of the design genius who comes up with solutions in a flash of insight. In this "eureka" moment, the story goes, the design solution appears fully formed in their head. But in practice, this rarely happens—the best designs are often the result of numerous rounds of exploration, feedback, and revision that help a designer understand the range of possible solutions to a problem.

This myth persists because others see only the finished product at the end of that long creative journey—they never see all the messy work that led to it. But the road to good design is filled with many twists, turns, and detours!

Have you ever opened your design software with a preconceived notion of how to solve the problem you're tackling? Of course you have; we all have. If you have an intuitive idea in mind, put a pin in it and keep exploring. I've learned from many years of designing and shipping products that intuition is often wrong. There are many reasons for this, but most often it has to do with the fact that your mental model of how people will behave doesn't match reality. Your intuitive idea may be where you start, but the only way to find the best solution is to explore many options.

- *Start as broad as possible*: I've found that it helps to make your design variations as extreme as possible in the early stages—don't worry, you can always rein them in later. What you're trying to do at this point is find the boundaries of your problem space. If you're not designing something that makes you a bit uncomfortable, you're probably not going far enough. (Just because you design it doesn't mean you have to show it to others.)

- *Visualize trade-offs in your design*: For example, consider the trade-off of density. You could make your design more dense, which would allow you to present more information up front at the cost of potentially overwhelming your users. Or you could make it less dense, which might feel clean but could detract from the experience because users won't see what they're looking for. Design both a very dense version and a very sparse version (and many versions in between) to see where they start to break down and where the design opportunities may be.

- *Get out of your headspace:* Show your design variations to potential users of your product, stakeholders at your company, and others who might be less familiar with your product. This is where it will help that your design concepts are extreme, as they will be much more likely to elicit a reaction (positive or negative) from people. Ask the questions that will help you understand what aspects of your designs are failing or succeeding.

After this exercise, you'll have a range of design directions, a bevy of feedback, and, most important, the valuable journey that you took to get here. You may not choose any of the designs you've produced along the way (after all, they were quite extreme), but this "throwaway" work often exposes the insights that lead to a breakthrough in the next round of designs, and to a final product that seems intuitive to the people who use it.

Design Thinking Workshops Will Change Your Process

Theo Johnson

Implementing design workshops in your UX process is an effective way to gain team alignment about users and help create a better product. Design thinking workshops not only help you conduct a thematic analysis of user research but also help focus the team on the user. Instead of the team jumping into detailed design with little to no insight, design workshops bring the team together to ideate on how to solve problems first. Design workshops are also a great platform for conveying summarized qualitative user research data, insights, and themes to your UX team early in the design and development of a project.

The first step in having a successful workshop is to plan ahead (approximately two to three weeks) and schedule the workshop with your team via calendar invite. You'll want to have designers, user researchers, engineers, data scientists, and other members of the stakeholder team in these workshops. It is imperative to have different perspectives of cross-functional team members for the design workshops since they are the platform to raise awareness about how your users think and react to the product you are building.

Prior to the workshop, make sure your team has access to all the research data, findings, and associated recommendations. It's also important to create instructions for your team, including what software you are using for the workshop. This way, the team can show up ready, with all of the necessary tools and data on hand for the workshop. There are lots of excellent online whiteboard tools available for design thinking workshops. If it's an in-person workshop, having a whiteboard and plenty of sticky notes on hand is a must!

Typically, I like to run 90-minute sessions for design workshops; these sessions are divided into four stages for exploring the nuances of the research findings:

1. User needs, likes, dislikes, and challenges (30 minutes)
2. Create themes and use cases for the user flow (30 minutes)
3. "How might we..." statements (20 minutes)
4. Reconvene and share thoughts (10 minutes)

This simple structure has been effective and efficient in summarizing and conveying research findings. In step 1 of the workshop, each team member can start adding sticky notes from the research data they had time to examine before the design workshop. Examples of user research data are things the team observed during the research, such as "users said they don't enjoy the checkout process" or "users expressed that they love the sign-in process." During step 2, the team figures out *why* users expressed frustration and encountered issues in their checkout process, or *why* users enjoyed the sign-in process. As you explore the different "whys," themes and patterns should emerge from the data that point you to potential solutions. The team can then produce use cases for user flow diagrams and identify where the challenges are during the user journey. In step 3, the team can utilize the "How might we..." statements methodology, which involves identifying opportunities for solutions and can be a jumping-off point for design. An example of a "How might we..." statement could be "How might we simplify the checkout process?" The team members end the session with step 4, in which they share their thoughts and opinions on the workshop and decide on next steps for the team.

Design workshops can help your team understand how and why users may be having difficulties with your products. They also bring stakeholders to a common understanding and align the team on a user-focused vision. A design workshop can help a team drive the decisions to create a functional and meaningful product that aligns business goals with the needs of users.

For more information about workshops, see Part III, *Align Your Team Around Customer Needs via Design Workshops*, page 81.

Visualize Requirements During a Workshop

Kristina Hoeppner

Spreadsheets are the tool of choice in the procurement process, but they are difficult to navigate during a workshop: horizontal scrolling makes it easy to lose focus among the dense information, and the font size is too small even when printed on large-format paper. This chapter suggests abandoning the spreadsheet and preparing your information in a more visual way so it can be consumed more easily during a conversation about project requirements.

Leave the Spreadsheet Behind

Inspired by the book *Gamestorming* (O'Reilly) by Dave Gray, Sunni Brown, and James Macanufo, and by previous success in reimagining a conference workshop by employing ideas from the book, I wanted to find a more engaging way to discuss software project requirements during a multihour workshop. The main goal was to show that the software already complied with a large number of requirements. Instead of arming myself with printouts of the spreadsheet, I copied each requirement onto a postcard and grouped the cards, and we focused on discussing items that needed additional insight.

On the day of the workshop, stakeholders were surprised our team had filled the table with the requirement cards instead of spreadsheet printouts. We proceeded to put all items into four categories:

- *Known*: Functionality is available and doesn't need discussion unless the client has questions. This was the largest group we wanted to demonstrate.

- *Configure*: Requirements we can meet by configuration or a simple customization.

- *Develop:* Functionality is not fully available, and we know how to achieve the desired result.

- *Investigate:* Items for discussion to clarify our understanding and determine a way forward.

By recategorizing the requirements in the spreadsheet away from their functional grouping, we could now focus immediately on those for which we had the most questions, shortening the meeting time by about 40%.

Prepare for and Run the Workshop

The aim: nobody needs to refer to a spreadsheet during the workshop. Choose categories for your requirements that support your goal (e.g., Known, Configure, Develop, Investigate), then assign a distinct color to each category and print each category name on a sheet of printer paper.

For the cards, set up a template in your presentation software using a regular printer paper format. Each card contains the following:

- The requirement's reference number
- A short title for the requirement in a large font
- A thick border around the reference number and title in the category color
- The requirement's priority, circled and put next to the title
- The requirement's text in a medium-sized font
- Your response to the requirement in a smaller font
- Questions you want to ask

Print four cards per page on sturdy paper and cut them to size to make them easy to pick up and handle in the workshop.

Before the workshop, group the cards on a large table by category and then group them thematically within each category. You can add sticky notes and jot down the themes if you have a lot of them.

During the workshop, discuss a single card or theme, take notes on the back of the card, and change the categorization if necessary, using the appropriate color for the new category. At the end of the workshop, sort all cards into their respective (new) categories again, and show the participants the stacks so they can see the progress made during the workshop.

Be Brave and Break Away from the Spreadsheet

Presenting dense information from a spreadsheet in a different way helps keep your workshop participants engaged. Participants can grab a card and read it or bring it into the discussion of another card, and at the end of the workshop they can see the progress that was made.

Put On Your InfoSec Hat to Improve Your Designs

Julie Meridian

You've finally created your design and fulfilled your dream of the experience your users will have. After you pause to bask in this ideal state, change your perspective: what might lead to an experience that doesn't look as good or act as well as you imagined? Take a cue from information security practices and start thinking like a hacker. *Information security* (InfoSec) is the practice of protecting information from risks. By putting on your InfoSec hat, you can anticipate solutions before they're needed.

When assessing your design, ask yourself these four questions—they are a good proxy for what may happen, accidentally or intentionally, in real use:

1. *Are you using only ideal content for the imagery and data in your designs?* Try images that are low resolution, too large, or too small so you can design smarter framing that keeps images of any type looking good. Whittle your activity feed down to just one post, or your infographic down to just one data point, to simulate an early or low-usage experience that might need more suggestions or prompts. Fill your product with similar content from different sources to see what's repetitive and should be bundled together. Anticipate these scenarios up front to design a good experience for all imagery and data, not just the ideal ones.

2. *Does your product need a reliable internet connection for some or all of its functionality?* Use (or simulate) a slower connection, or shut off your connectivity altogether. Better yet, simulate what happens if your connection drops in the middle of an important task, and what happens if it returns later. Work with your developers on solutions to avoid data loss and on messaging to accurately describe what's going on and how to recover from it.

3. *Is it usable in other situations?* The way people interact with the world varies, and you don't need to be an expert on accessibility or internationalization to consider those differences. For the display, try using a

smaller window to emulate a mobile device or a scaled-up screen. Make sure your design reacts appropriately to visual accommodations, such as increasing the text size (the operating system setting for text scaling), or speaking the page hierarchy and visual descriptions out loud (screen readers). Imagine you have difficulty reading, and use the most prominent words to figure out what to do. Imagine any text doubled in length or written right-to-left to gauge how it will react to different languages. Once you've explored all these views of your design, see how far you can navigate while restricting yourself to a single input (just keyboard, cursor, touch, or voice).

4. *How might your users trick each other?* Once you've exhausted these methods, get to the heart of the InfoSec mindset and imagine the sneakier ways your product may be disrupted. If there are technological and social ways to exploit your product, seek to mitigate them in advance. How could you convince someone to take action or give you information you shouldn't have? How could you misrepresent yourself? Every person who can access your system could be a target of social engineering. Do your best to prevent manipulation by bad actors, *before* that manipulation occurs.

All tech is designed with inner logic that designers should understand and assume users may exploit. Although we designers have the best intentions, not all users will.

This InfoSec approach can challenge any system, especially products with potential for real loss and abuse through user-generated content and multi-user interactions or in sensitive industries like healthcare or finance. That's why you should put on your InfoSec hat and hack your designs up front to spot subpar experiences and security exploits before your users do.

On-Brand Whimsy Can Differentiate Your Mobile App

Martha Valenta

When designing an intuitive experience, it's obviously important for us to use common, familiar interaction patterns and to use them consistently throughout an app or a website. At many companies, a design system will be available, and keeping consistent with the design system will keep your systems intuitive.

But sometimes there is a good case for dreaming up a completely new interaction that suits the specific app you're designing. It might help users transition from the physical world to how it's done in the app, such as the Weather app showing motion imagery of the current weather. Or it may simply bring a special on-brand whimsy, such as the slot-machine-styled restaurant picker in the old Urbanspoon app.

When adding delight, how do you come up with this special on-brand interaction while keeping the design intuitive? Here is a procedure for brainstorming and refining unique interactions:

1. *List every possible experience*: Think broadly in this step. Review any prior user research so that you have a rich pool of information to pull from. Then make a list of users' experiences. Think about interactions pertaining to the task at hand and delightful experiences these users typically enjoy outside of the app. Once you have a list of user-focused experiences, start thinking about the business. Consider feelings or experiences that the business or marketing stakeholders would like the user to have and add them to the list.

2. *Select an experience*: Circle experiences that seem to be perfect for the app. Put a star next to the strongest experience. Cross out the

experiences that won't be delightful or that don't make sense. Remember that we want only ONE unique interaction in the end.

3. *Brainstorm:* How might we incorporate this experience into the mobile app? Create several quick prototypes of the experience's unique interaction. If possible, bring other team members into this brainstorming session. If the experience you chose isn't working out, pull another one from your list.

4. *Leverage mobile inputs and outputs:* As you prototype, consider how the various inputs and outputs of the mobile device might play into this interaction you're prototyping. For example, a calendar app might pop open a birthday cake with candles on your birthday. Take it one step further and let the user blow out the candles (via the microphone input). Here's a list of possible mobile device inputs and outputs: microphone, speaker, camera, fingerprint biometrics, touch screen, GPS, accelerometer, barometer, altimeter, WiFi, Bluetooth, flash/flashlight, and vibration.

5. *Ask a developer:* Discuss the best result of your brainstorm with your developers. Run it by a developer who is new to the company and motivated. Then run it by a developer who has been with the company for a long time. Balance the feedback they give you to determine a ballpark estimate of how much effort it might take to bring this interaction to life. Don't be discouraged if the developers tell you that it would take too much effort.

6. *Share the idea:* It's time to share your idea and the roughly estimated effort with stakeholders. It's best to do this in person or via a video meeting. Create a one-page overview of the interaction with the ballpark estimate from the developers. Show your interactive prototype during the meeting to help others envision it. Capture their feedback. It's okay if they aren't excited about it. Use the feedback to create the next iteration. It could also give you some insights into how the company or those stakeholders prefer to work. Worst-case scenario: if they don't use the idea, you can write a case study around it and put it into your portfolio.

Remember, not every interaction should be a fun interaction. Most should be standard and simple. But including one optional, delightful, brand-enhancing experience in an app can be a differentiator that makes your company's app the one that users prefer.

Don't Perform a Competitive Analysis Before Ideating

William Ntim

The practice of performing competitive analysis before putting your ideas to paper, ideating, or concepting will create cognitive bias during your creative process. We're often quick to check what competitors are doing as soon as we are presented with a problem statement, but this may be a creativity-limiting step.

Starting your design project with a competitive analysis can anchor your creativity, and copying a competitor is not an effective way to lead the herd. This can cause poor UX to proliferate if ineffective UX is repeated and becomes the norm. You probably have heard this before: "Amazon does it this way," "Airbnb does it this way," and so on. The key part people forget is that Amazon and Airbnb created their designs for a specific reason. Lots of research went into arriving at their current experience. Companies may have similar audiences, but placing *your* user at the center of your design decisions is how you meet your user's needs. The other reality is that these "competitors" are already six months ahead with whatever feature you're looking to copy, and it will probably be another six months before your team is ready to roll out said feature. By that time, the competitors are probably going to be rolling out another new feature. Yes, this is the cycle, and it can continue indefinitely if design teams do not begin and end with their user.

There may, however, be rare scenarios in which a product's entire goal or positioning is to surpass the competition. Even then, this analysis should be performed in a way that explores competitive gaps and pain points. This strategy is referred to as blue ocean strategy, as coined by W. Chan Kim and Renée Mauborgne in their 2004 book *Blue Ocean Strategy* (Harvard Business Review Press). This is where the simultaneous pursuit of differentiation and low cost will open a new market space and create new demand, and it is an appropriate time to conduct a competitive analysis at the onset.

When you are presented with a problem statement, gather all the research you can and obtain a thorough understanding of the problem in relation to *your* users. Understanding the problem statement and the user persona or group you're designing for is the first part of a successful process. Then proceed to sketch, ideate, and concept solutions based on that data. After you have explored a few options, it's safe to perform a competitive analysis to inform what you have designed.

In summary, when you are presented with a problem, you should do the following:

- *Seek to understand the problem statement*: Ask questions about why the solution was requested, who requested it, how the problem was discovered, its impact on the experience and business, who the customer is, and so on. Remember, your goal here is to obtain a thorough understanding of the scope of the problem and its impact on the business and end user.

- *Dig into any available research*: There may or may not be user research data available for review. If there is, thoroughly review this data, as it could contain useful information to help you understand your users or landscape better.

- *Harvest your ideas first*: After gaining a thorough understanding, you are ready to sketch, ideate, and concept solutions. This step is where you'll tap into your creativity and retrieve original solutions before opening yourself up to the external bias of competitive analysis.

It is nearly impossible to completely avoid cognitive bias, but it is important to reduce any biases that may restrict your originality and creativity. You are a UX professional because of your skill and ability to create. Put that to work first before you open the door to external influences.

PART IV

Content

Design for Content First

Marli Mesibov

How do you start a design? Some designers start by identifying the goal. Others start by blocking out sections of a typical home screen. There are many approaches, but *content-first design* will guarantee that you consider what your end user most needs.

Content-first design is an approach to UX that begins with, well, content. More specifically, this process connects design elements to the exact information your end user needs and wants. In some ways, content-first design is a mindset, but it is also a methodology, with steps that anyone working in UX can follow to improve their work.

The Content-First Mindset

When content comes last, projects run late and budgets run high. It happens often—the project is "complete," and then it turns out that none of the existing content works with the new design templates. So either content needs to be rewritten or designs need to be amended. From this perspective, content-first is a mindset. It's a way of looking at a need—be it a transactional portal or a library of articles—and considering the *types of content* as part of the requirements.

For example, in a digital library the articles might be intended for several different audiences. If you review the articles and find that out, you'll know that creating a tagging system is a requirement of the new digital library. Perhaps some but not all of the articles include images—another requirement, then, will be flexible templates. Maybe the target audience responded well to concepts that included a home page with six categories of content. If there aren't any existing articles that fit some of those categories, the project scope needs to change to include writing more articles (or alternatively, it was a mistake to test a concept that wouldn't work with the existing content).

Put another way, content-first means reviewing the intended content before beginning design. Having a familiarity with *what* the target audience is taking in will have a significant and positive impact on the project.

Content-First as a Methodology

As already stated, content-first is also a methodology, with a set of steps any team can follow. When teams follow a specific content-first process, it may include the following steps:

1. Identify the various audience segments and their top needs.

2. Prioritize those needs: assign each one a number to indicate its relative importance. No two items can have the same number.

3. Before designing a page, write a concise message (one to two sentences) that identifies what the page should get across.

4. Select the needs this page will fulfill. Compare them to the message to make sure the message and the needs are in sync.

5. Design the page to further support the message and these needs.

As a methodology, content-first ensures a page is intentional and intuitive. When you design with a concise, meaningful message in mind, you won't be tempted to incorporate too much into a single screen, or to try to make multiple areas the "most" important. This clarity will help your target audience accomplish their goals.

Speak to Your Audience

Ultimately, the goal of a design is to move your audience forward on a path. A good design evokes a message and provides information, whether with words, icons, or design elements. With that in mind, content-first design is the only way to find your message first and then use your design to communicate it.

Align Your Tone, Voice, and Audiences

Marino Ivo Lopes Fernandes

Who do you think you are?

Who do you think you're talking to?

You probably had clear reactions to those questions. Maybe you were confused or annoyed or stopped reading. These questions illustrate that tone and voice manage the relationships you earn with your readers. You can learn dos and don'ts about tone and voice on the internet: how personal pronouns, punctuation, and emojis help you manage your writing style. But tone and voice are not gimmicks or simple tricks to sound friendly. Managing tone and voice depends on a writer's answer to three who questions (and not the snarky ones you just read):

- Who are you addressing?
- Who do you want readers to think you are?
- How do you represent who your readers are in your content?

There is a difference between writer-based prose and reader-based prose. Tone and voice moderate how our readers interact with our content. In *reader-based* prose, writers do the hard work for the reader: they consider the needs and interests of readers. It's an act of care. *Writer-based* prose follows a stream of consciousness, is ego-driven, and assumes that readers can access the writer's mind. Such a relationship with a reader is unearned. If the reader would benefit from follow-up questions to understand you, you're engaging in writer-based prose.

Here's a scenario of these ideas in action: many writers agonize about the perfect words for sentences. That's all well and good, but I want to convince you to think about balancing the words you could choose and the words available to your readers. Consider the most 2020 sentences ever:

> Please utilize the following survey to determine your vaccination eligibility. This tool will be used for the sole purpose of determining eligibility for receiving a COVID-19 vaccination.

Everything I'm trying to tell you about tone, voice, and audiences is not informing these sentences. These sentences assume that the reader and the writer went to the same meeting. Now consider an alternative sentence:

> You can use this tool to determine eligibility for vaccination.

This is better. It recognizes that a person is taking an action on the other side of the screen. It chooses a friendlier and shorter verb (verbs are where it's all at). But the nominalizations in this sentence (eligibility and vaccination) require another verb for the reader to understand. Finally, look at this final revision:

> Please use this tool to see if you can get the vaccine now.

This is the best choice because it embraces more audiences—it assumes the reader missed that meeting. It's friendlier because there is less to figure out: yes, there could be some ambiguity in what we mean by *can get the vaccine* that is cleared up by *determine eligibility*, but a word can't be clearer if readers don't understand it.

If you're doing it right, your audience won't think about tone and voice; they'll feel like they got what they came for:

- Do the hard work. Make the content accessible to your readers. Assume readers are busy. Don't expect them to persist through your writing at all costs.
- Consider who is actually reading your content and what they want from it.
- Choose easier words when possible. Balance the tension between the perfect word and words more people can understand with less effort.
- Vary sentence length, but keep sentences short; 15 to 20 words per sentence is friendliest.
- Keep the verb and noun close together and as close to the beginning of the sentence as possible.

Managing voice and tone is not a matter of linguistic trickery. It's about delivering a message appropriate to the occasion and to the needs of the audience.

Mind Your Error Messages

Jennifer Aldrich

Every decision that designers make has an impact on users. Even the tiniest details, if not handled correctly, can create a negative user experience. One of the most fascinating but often neglected aspects of design is crafting error messages. The psychological response people have to poorly crafted error messages can be extremely jarring.

Prior to joining the UX industry, I was a software trainer and had the opportunity to travel across the US, observing user behavior and identifying unintuitive areas of our product line. Some products did an excellent job calmly alerting users that something unexpected had occurred. They had a friendly, professional voice and kept a consistent tone. Other products threw hideous, scary-looking error messages. The symbols, the wording, the color choices—really, every aspect of the messages induced stress.

At one point during my travels, I observed two users working in different pieces of software that errored out. One product displayed a giant red X with the word ERROR in all caps and code in a huge font beneath it. The user gasped, closed the browser, and shot back in his chair.

The other user received a message that read, "Something just happened on our end, sorry about that. Please refresh your screen and try again." The error code was listed in small text below the message, along with an option to expand for additional detail. The user calmly refreshed the screen and continued editing.

The impact that friendly error messages can have on UX is incredible. The actual errors these two users encountered were identical, but how the different softwares handled them was *vastly* different.

And error messages aren't just an annoyance—studies have shown that when a system breaks down, it can raise cortisol levels, a biomarker of physical stress. Your error messages can legitimately impact the health of your users by increasing their stress when errors occur.

So how do you create a quality experience? Make sure the voice and tone align with your brand and target personas, and avoid the use of jargon and technical language if it's not appropriate for your audience. Always conduct research to test messaging and to determine if the language is clear. See Part IV, *Align Your Tone, Voice, and Audiences*, page 128 for more information about tone and voice.

For the product I described earlier, the customer base consisted primarily of teachers. We made our error messages concise and friendly and applied progressive disclosure of detail. Each error displayed a small illustration with a one-sentence explanation of what had happened and a call to action. Beneath was an expandable arrow that contained all of the details that our target persona didn't need to see 99% of the time.

Know your audience. If developers are using your product and a missing semicolon can break a program, they may prefer error messages that immediately display all the details they need to take action and resolve the issue without any extra clicks. In their case, hiding details would actually detract from the UX rather than improving it.

To craft a high-quality error message, explain the error, provide technical details if needed, and give instructions on how to resolve the issue. Maintaining your navigation structure when an error occurs is a great way to keep users calm. Remaining in the framework of the website makes the shift to the error screen feel more natural and less jarring.

There are millions of error messages thrown by software and websites each day. When you're designing, don't forget to pay attention to the "little big" details, like humanizing your error messages while keeping your target personas in mind. It will make a huge difference in UX and product stickiness.

A Shared Vocabulary Can Increase Team Efficiency

Matthias Feit

A colleague and I once famously spent more than half an hour discussing the term *copy deck*. It was part of a project briefing, and we had very different opinions about what the term means. Heated arguments and a massive undertaking in information retrieval ensued. It wasn't exactly time well spent, and the argument most likely could have been easily avoided if we had a common glossary.

When you are collaborating with people from different domains and departments on a project, a shared understanding of words and their meaning is key to preventing frustrating and time-consuming mistakes.

Coming to Terms with Terminology

How many times have you been in a meeting in which two people are talking about the same thing but are confusing each other with slightly different terminologies? Or even worse, with very different meanings for the same word?

Language is often messy. Take a minute and think about the different words you have seen people use to describe a UI element that is used to display different images over time:

- Carousel
- Image slider
- Animated gallery
- Hero images

This is just one example of somewhat abstract terms that can be explained to us in very different ways. These abstract terms describe processes, roles we assume in these processes, design deliverables, and tools or methods we work with.

If you have a couple of hours to spare, try to determine the difference between *UX* and *CX*, or try to find a definitive definition of *Jobs to Be Done*. You get the point—you can't.

A team that lacks a shared vocabulary is more likely to waste time and resources due to misunderstandings about what means what. This is even more likely to happen with remote teams and people who are not communicating in their native language. Luckily, there are things you can do to prevent this from happening.

Record and Clarify Ambiguous Terms

People often don't like to admit not knowing the meaning of technical terms used by colleagues or clients. Not understanding what the other party is talking about could fuel the fear of appearing incompetent.

On the contrary, asking a colleague to define what they mean when they mention *content strategy*, for example, shows that you paid attention and really want to understand them. It is entirely possible that you also clarified the term for others in the room who were also afraid to ask.

Create a Glossary

Your team should have a unified language when communicating with clients. If you find that there are a multitude of ambiguous technical terms and abbreviations floating around, it's useful to sort them out once and for all. This can be done in a workshop where you:

- Write down the terms in question.
- Collect and discuss people's definitions.
- Find consensus on which definitions are accepted.
- Add the definitions to a glossary that is accessible to everyone.

A simple list that is shared throughout your organization is a great start. Make sure that it is up to date and that everyone is kept informed about new revisions. You could even use this exercise as a catalyst for creating a *UX playbook*—a document that lists and explains your tools and methods—to help onboard coworkers.

Simplify Your Language

We have to constantly remind ourselves that the language used in UX processes (think *heuristic evaluation, think-aloud method, How Might We's*) is not necessarily accessible to other coworkers. We need to clearly make our points using words that everyone in the room understands. A shared glossary goes a long way toward making that happen.

Break Your Lorem Ipsum Habit: Sketch with Words!

Emily Roche

It's the beginning of a project. You're in UX heaven as you sketch out ideas and interactions that will shape the experience. You sketch with shapes and icons to indicate where images and videos will go, and add *Lorem Ipsum* to show where the words will go.

WAIT. On behalf of UX writers and content strategists everywhere, please don't do this. A sea of *Lorem Ipsum* indicates that the content we need to write is an afterthought in the design. No one's really thought about what the words will describe, or what people should do when they see CTA (call to action) copy in a text link or button.

Why Lorem Ipsum Doesn't Help

It's wonderful to see UX writers and content strategists included as key members of digital project teams. While UX writers are becoming more common, they're often in short supply. Organizations tend to hire more "traditional" UX talent, such as designers, rather than writers. As a result, UX writers usually juggle several deadlines at the same time.

Also, UX writers typically don't start working until it's time to "add the real words," which is usually well after the project kickoff date. This means that when a UX writer joins the team they're already behind. Why? Because even though *Lorem Ipsum* shows where words will go, it doesn't convey the meaning behind them. UX writers will need to ask myriad questions to fully understand the intent behind all that mocked-up text, which can add to already-compressed project timelines.

This is where you, folks who are not UX writers, can help by sketching with words.

Here's How to Sketch with Words

Think of words as design elements with jobs. They can educate, entertain, or encourage someone to do something—the possibilities are endless. The next time you're working on a design and reach a place that needs placeholder copy, jot down some words that quickly describe what the copy is all about. That's it! You're sketching with words. Here are some examples of sketch copy and what each conveys:

- "This headline has six words total" (conveys word count)
- "This subhead describes the product benefits" (conveys intent)
- "CTA is 'Watch the video'" (conveys action and the content type users will see)

Sketching with words will help you design a more holistic experience from the very beginning. When you use real words from the start, everyone on the team can see what's going on in the designs and how the interactions should come to life.

It will also surface any surprises or weaknesses in the design and give you more time to fix them.

Stronger Starts and Smoother Finishes

Recently I was a UX writer and equal partner with a designer on a web redesign project. We wanted to include teaser copy in a product module and made a space for it in the Figma design sketch. As soon as I started writing, I realized the client didn't have the information we needed. The copy felt empty and full of "fluff," so we removed it from the design. This let me focus on writing stronger headlines and gave my design partner more space to showcase product visuals. The result was a smarter design and a happy client.

This is just one example of how sketching with words can help a project run more smoothly. Please don't think your sketch copy isn't good enough, or that the UX writer thinks you're crossing onto their turf. When they see you've used real words to create a solid foundation for them to write from, they'll be at a loss for words to thank you. Happy sketching!

PART V

Research

Always Go for the Why— the Immutable Basis of Great Design

Andy Knight

At the beginning of any design process—of any creative process, really—you must relentlessly ask, "Why?" *Why* is the key to all decisions and all paths followed or ignored. *How* changes constantly, and it should, because you will learn new things throughout the process. But if every member of the team understands *why*, you have a fighting chance of getting the *how* right.

When dealing with business partners and sponsors, particularly those who've never been through an end-to-end design process, you must be especially relentless in your pursuit of *why*, because stakeholders often come with preconceived notions of *how*. Many times, I've framed the *why* question as, "Help me understand your business goals and how you're going to measure success." And more times than I care to remember, the response has been, "Why do you need to know that? You're just doing the design. Here, I did this over the weekend in PowerPoint, based on what our competitors do."

I generally don't resort to salty language or worse. Generally. Instead, I take a deep breath and explain: "If I understand why this is important to your business and your customer, I can ensure that we're truly solving your business problems and customer needs." For partners, framing the question in terms of what behavior they'd like to change often helps. To help them visualize how success should be measured, ask them to imagine it's the end of the year, and they want to convince their boss they deserve a large bonus: in what areas would they like to prove they materially moved the needle? Get them talking in detail about outcomes, and you'll often get to the underlying *why* quickly.

With customers, it's much the same game. Understanding why a user comes to an experience in the first place is at the core of empathy—and getting to the *why* with users is the most fascinating part of design. In user-centered design, we deploy many means to understand users' needs and intentions, from one-on-one usability studies to focus groups to large-scale data analysis of existing use patterns. It's critical to remember that customers often don't fully understand why they are doing what they're doing, and we often lack the knowledge and context to ask the right questions to get there. This realization is at the very root of design thinking: by spending time closely observing users (or observing how they actually use existing designs) and asking "why" repeatedly, we have a fighting chance of seeing things through their eyes and feeling what they feel. It's critical to remember, though, that the *how* can often obscure the *why*.

Recently, my team worked to figure out why a large percentage of users dropped out on the second step of a complex process. The easy assumption was that the step was too difficult, but after observation, we recognized a pattern: a lot of users were exiting and then never completing the transaction via any channel. Our hypothesis: people were just "shopping" to see what was available through the process and had no intention of actually completing the task. Their *why* wasn't directly addressed elsewhere in our experience, so users found a work-around. To satisfy this newly understood need, we built an alternative path that answered their question more simply and directly. The result: real-task completion percentage shot up, *and* customer satisfaction improved.

The best part about this "design" process was that it integrated the full team: business, digital, and technology partners were all invested in figuring out why this was happening, and in coming up with a solution that solved a problem we didn't know we had.

The Participant's Well-Being Is Your Responsibility

Danielle Cooley

In our quest for good research observations, we sometimes forget that the participant's well-being is more important than our research goals. Paying attention to the participant's real-time needs can enhance research outcomes because you're getting the participant's whole self, and this responsibility belongs to the researcher.

Pay Attention to Physical Needs

The researcher's responsibility for a participant's physical needs is most applicable in in-person sessions. Ensure the study environment is at an appropriate temperature and that comfortable seating is provided. Keep noises and distractions to a minimum. Offer the participant some water or a snack—a "hangry" participant isn't going to give you the clearest commentary.

Physical needs can also extend into actual safety issues. In a summer 2015 study for an automotive manufacturer, we were evaluating a completely reimagined dashboard display and center console. These sessions were conducted inside a parked car. Interior temperature was a major safety concern. On a hot day, the temperatures could exceed what is reasonable or safe. To accommodate this reality, we scheduled breaks (to air out the car), ran the air conditioner on full blast between sessions (alas, not during, as it shorted out the fragile prototype and made it hard to hear the participant), cracked windows (trying to maximize airflow while minimizing recording noise), and ensured everyone involved had water on hand.

Mental and Emotional Factors Affect the Research, Too

Participants' emotional needs are also important. First, a participant who doesn't feel safe isn't going to provide candid commentary—they're far more likely to tell you what they think you want to hear. This is one reason Rolf Molich et al. determined that "building trust and rapport" is a key part of usability test moderation.[1]

As with physical needs, attention to mental and emotional needs can also veer from participant comfort into participant safety, such as in the case of evaluation of products associated with trauma or severe health issues. In one instance, the product being evaluated was for newly diagnosed cancer patients, and the project team felt the discussion would be too upsetting for people with such fresh trauma. Instead, we recruited participants at least six months out from their diagnosis. Similarly, a recent study with chronic obstructive pulmonary disease (COPD) patients required very careful moderation. Several participants reported "just trying not to think about it too much," when of course that's exactly what we were asking them to do.

Don't Be Afraid to Stop the Session If Necessary

There are scenarios in which you can't accommodate the participant's needs *and* continue with the session. A few years ago, at the start of a remote, moderated session, the participant said, "I think I might have to reschedule. I'm having chest pains and think I should go lie down." I told her to hang up and call 911. She could not be persuaded to do so, but we certainly stopped the session (she rescheduled and was fine). This incident illustrates just one example of how sessions can go very awry. Donna Tedesco and Fiona Tranquada discuss several extreme scenarios in *The Moderator's Survival Guide* (Morgan Kaufmann).

---

1 Rolf Molich et al., "How Professionals Moderate Usability Tests," *Journal of Usability Studies* 15, no. 4 (2020): 184–209. https://oreil.ly/XYn7A.

Caring for the Participant Is in Everyone's Best Interest

Keeping the participant's physical and emotional comfort and safety in mind is paramount—not only because it's morally and ethically important but also because you get better data this way. Participants who feel stressed or uncomfortable (or worse, unsafe) aren't likely to share their candid thoughts about whatever it is you're researching.

UX practitioners are here to make people's lives better and easier—and that includes the lives of our research participants.

Diverse Participant Recruiting Is Critical to Authentic User Research

Megan Campos

When I began my career in UX research, I noticed that all of our participant recruitment screeners called for "a mix" of criteria like race, gender, household income, and education level. I also noticed that asking for "a mix" led to demographically homogeneous participants. The team always came up with reasons why we could not or would not push for a more diverse recruit: usually the reason was that it did not seem like demographics had a direct impact on the experiences and behaviors we were studying.

But individual mixes of demographic criteria *do* have an impact on how we experience the world and on how the world experiences us. Everything from the kind of healthcare you can expect to receive to the acknowledgment of your basic rights as a human being can be shaped by your demographic makeup. To argue that demographics do not shape behavior is to ignore all evidence to the contrary; to acknowledge that demographics do shape behavior is to render moot the argument that they do not matter in user research.

Once I acknowledged that asking for "a mix" wasn't cutting it and was likely resulting in a less-than-full picture of our user groups, a different approach became necessary. It was time to get specific:

- *Flesh out the demographics:* We examined the client's existing population to gather data on its demographic makeup and proportion; if the client didn't have that data, I looked at US Census data (freely available online) to determine the desired proportion of the demographic constituencies we wanted to recruit.

- *Establish margins:* The screener called for either a specific number ("5 Black/African American") or a specific range ("3–5 Black/African American") for all criteria where demographics were critical to fully understanding the audience. This essentially abolished a vague "mix" from our screener vocabulary.

- *Make the argument:* Unless you're running a study by yourself, you will need to bring your team and/or your stakeholders on board with your approach. Spend some time researching the demographic breakdown you are asking for and establish a clear rationale.

- *Remain firm:* There is a high likelihood that you will get pushback from your recruiter; before you sign a contract, make sure the recruiter is on board with what you are asking. You are hiring them to find the people you are looking for, and those people definitely exist.

- *Establish a feedback loop:* Once the study is complete, follow up with your recruiter to discuss how they did or did not meet recruitment expectations, including the degree to which you needed to insist on certain criteria. Constructive feedback is mutually beneficial: they will learn how to evolve their practice and retain your business, and you will see subsequent recruits that better align with your target audience.

This approach has worked incredibly well in recent studies. The findings we've uncovered as a result have painted a far clearer picture of the target audience, including important experiential and behavioral differences along demographic lines. Uncovering these kinds of findings has allowed our team to come up with solutions to problems that otherwise might have gone unnoticed or ignored. In most cases, we also found that when we took into account the needs of a typically misrepresented group, the solutions we came up with were likely to make the experience better for the broader target audience.

We should be doing the work before the study to understand the demographic makeup of the audience and then be insistent that the recruitment process yields the people we need to talk to, even if they are a little trickier to find.

Build a Culturally Reflexive Professional Framework

Monet Burse Moutinho

> *"What we choose to design and, more importantly, what we choose not to design and, even more importantly, who we exclude from the design process—these are all political acts."*
> —*Mike Monteiro,* Ruined by Design: How Designers Destroyed the World, and What We Can Do to Fix It *(Mule Books)*

There are many skills that every designer or researcher ought to possess, but one of the most important is reflexivity. *Reflexivity* is a critical examination of one's own beliefs, judgments, and practices during the research and design process and of how they have influenced the end research and/or design product. This concept is often used in the social sciences with the intent to address objectivity in research practice. Reflexivity is the conceptual framework upon which my five-question guide will be built.

As a lifelong student of anthropology, I do not believe that research or design is objective or benign. Because of this, it is imperative that designers and researchers make a plan to address the consequences of our designs and research artifacts. Design and research are purposeful events that at best can make our life easier but at worst act as a vehicle to drive inequality, reinforce cultural stereotypes, and proliferate bias. Creating a battle plan to address the negative implications of design and research should be a requirement for every researcher and designer. Such plans should not be a final step in the product development process but instead should act as a guide to drive inclusion, examine bias, and encourage reflexiveness.

The following considerations should be made during the *entire* research and design process:

1. Who does my design and research include or exclude?
2. Where do my personal beliefs diverge from or converge with my research or design?
3. Am I a member of the community that my research or design intends to impact?
4. Does my research or design have any cultural implications? Have I considered those?
5. Is my team socioeconomically homogeneous in some way (for example, from the aspect of race, ethnicity, education, or economics)? If it *is* homogeneous, is it possible to diversify?

By answering these five questions, you can begin to inject an ethics of care into your work. The questions can also act as a vehicle to drive conversations in your organization around diversity. They are intentionally abstract to provide flexibility in multiple work environments. It is important that designers and researchers critique themselves before the output of design or research artifacts. A critical examination of how and where one's deeply held beliefs, attitudes, and concepts intersect with a product build is not only responsible and ethical but also something that every UX designer should know how to perform.

Know These Warning Signs of Information Architecture Problems

Kathi Kaiser

When someone is struggling to use a digital product, it's often easy for UX researchers to see why: the screen may be too busy, the wrong action might be highlighted, or the copy may be unclear. When the problem is a flawed information architecture (IA), however, the warning signs can be more elusive. *Information architecture* refers to how content is organized into categories and how those categories are labeled (for more information, see Part III, *Don't Forget About Information Architecture*, page 40). Problems caused by a confusing or incomplete IA impact the user's experience holistically; they show up in people's behavior indirectly, making them more difficult to observe and diagnose.

Here are five behaviors to watch for in usability research that signal problems with the product's information architecture:

1. *Driving in circles:* In the real world, when people get lost, they look to signs and landmarks to help them recover. In confusing places (ever drive in Boston or in London?), these strategies can fail, and people may find themselves literally driving in circles. This happens in the virtual world too: when people inadvertently visit the same page multiple times, via multiple routes, they haven't formed a mental map of how the content is organized. This prevents them from moving through the product efficiently and completing their tasks.

2. *Navigating in place:* Have you ever watched someone repeatedly click a link to the page they are already on? This simple yet baffling behavior suggests issues with IA and navigation. First, the user doesn't recognize where they are—they aren't oriented to their location within the overall structure. Second, the link label wrongly implies that the subsequent screen will offer something different than the current screen. The label

could be faulty, or the content itself could be missing the mark. If you observe people failing to progress, it's a sign that the paths through your structure are not sufficiently clear.

3. *Heading for home:* When participants abandon their chosen path and head back to the home screen to try again, they're telling you it's easier to give up and start over than to reorient themselves with your navigation system. Heading home is a sign that users have lost faith in their own ability to get unlost. Sadly, they are more likely to blame themselves than realize the problem is with the product's IA.

4. *Getting half credit:* When participants declare they have completed a task but haven't found the intended screen, the culprit is often how the content is organized. People expect related information to be consolidated. Once they find some information on their topic of interest, they are quick to assume that they have found all of the relevant content, so they stop looking. It doesn't occur to them that additional information might be available elsewhere in the same product.

5. *Bet you can't do that again!* If participants struggle to complete a task before ultimately succeeding, there's a handy trick to explore the role that IA issues may have played in their troubles: ask them to do it again. After a short break or distraction, if they can't easily show you what they've already done, they haven't internalized the product's structure. The closer the interface aligns with the way users think about the domain, the easier it will be for them to learn it for future use.

If you notice any of these behaviors during usability research, don't despair! You're a step closer to knowing why participants are struggling. Since IA issues can lead to confusion and frustration throughout an experience, it's critical for UX researchers to notice these warning signs early and adjust the design to better meet users' needs.

Bring Themes to Exploratory Research

Shanti Kanhai

Exploratory research is an exciting stage of a project with many unknowns and new insights to discover. Different from research at later stages of the project, *exploratory research* requires a more flexible approach to interviewing to allow the conversation to stray from the interview guide. Unstructured interviewing using themes keeps the level of moderator control low and gives participants the freedom to show you what is important to them about the topic at hand.

Degrees of Control

Interview methods differ in the amount of control the moderator exercises over the interview. They range from a low level of control in unstructured interviewing to a high level of control in structured interviewing, with varying degrees of control in between. A popular interview method is to bring a list of standardized questions to the research session. This form of structured interviewing gives the moderator a considerable amount of control over people's responses by providing the same stimuli to each participant. The benefit of controlling the input is that you can reliably compare output between participants. The downside is that it restricts the range of potential responses from each participant.

In exploratory research, you can keep the amount of control to a minimum by sticking to the unstructured interview method. In this type of interview, you keep the conversation focused on a theme while giving the participant room to define the content of the discussion. This approach helps you uncover new topics of interest that might have been overlooked. Anthropologist H. Russell Bernard summarizes it this way in the book *Research Methods in Anthropology* (AltaMira Press): "The rule is: Get people on a topic of interest and get out of the way. Let the informant provide information that he or she thinks is important."

The Power of Themes

In qualitative research and analysis, themes are used as high-level abstractions to structure findings in research data. Examples of themes are *financial management strategies* or *social media behavior*. Working with themes has several advantages:

- By keeping the interview unstructured, themes provide a playground for *exploration and experimentation.*

- Themes help understand the *high-level picture* without getting lost in details.

- Themes serve as a *discussion guide* to ensure the interview stays on topic. Whereas a question-and-answer discussion dictates every step of the interview, themes act as a framework for eliciting stories. They provide structure while at the same time keeping the tone conversational.

Define Your Themes

An effective way to define your themes is through an ideation session with your team to discuss the direction of the project. Brainstorm with sticky notes, and group overlapping ideas to discover the main themes. Then refine the themes using these guidelines:

- Use four to six themes for a single research session.

- Keep the themes *short* so you can quickly refer back to them during the research session.

- Keep your themes *neutral.* Do not reflect opinions or assumptions in your themes, as they may steer your participant's responses.

- Use your themes as the foundation of a *mind map* you can use to take notes during the research session.

In exploratory research, aim for a low degree of moderator control to allow the participant to teach you what is important. Themes let the participant teach you and make you the apprentice within the boundaries of predefined topics. Who knows what unexpected insights you might uncover?

Embrace Your Ignorance

Jon Robinson

In his book *Astrophysics for People in a Hurry* (W. W. Norton), Neil deGrasse Tyson writes, "Ignorance is the natural state of mind for a research scientist." This quote instantly struck a chord when I first read it, and although I've butchered the exact wording many times in front of an audience, it's an idea that resonated during a turning point in my career.

When I was a young designer, I thought I knew everything. Before making the jump to UX research and design, I spent more than 10 years working for creative agencies, a community that often relies on knowing—or claiming to know—what an audience wants. But the older I got, and the more experience points I collected, I found discomfort in the confidence of knowing.

Placing too much faith in what I knew only led to inflexibility. Too many failed experiments taught me that assumptions are nothing more than faulty knowledge, but being aware of the knowledge I lacked allowed me to make better, more informed decisions. How? By ignoring my pride and not being afraid to ask any question, no matter how obvious it seemed.

If you want to be a better problem solver, then you need to be more focused on learning than on demonstrating the knowledge you already possess. You can't be afraid to ask stupid questions and challenge what others "know." When you hide behind your ignorance—pretending to know rather than challenging ideas—you fail to contribute and uncover real problems. Rather, embracing a curious, question-based approach never fails to support informed decisions, while also putting others in a position to more effectively contribute their own expertise.

UX research is a science, and science begins by recognizing that we don't know something. What device do our users prefer? How do they feel when using our product? These aren't questions to be debated. They require us to practice constructive ignorance—to focus on expanding our understanding by embracing a learner mindset, rolling up our sleeves (or throwing on our lab coats), and setting out to uncover new knowledge.

Here are a few actionable ways you can support that:

- Find the courage to explore things you don't know.
- Understand that the goal of an experiment isn't to predict the outcome, but to answer that which you can't predict.
- Instead of validating assumptions, challenge them.

I'm sure you've been in a situation in which you were afraid of—or discouraged from—asking questions. Next time, don't leave those questions at the door. Embrace the desire to know what you don't know and ask those questions. All of them. That's how we really learn to empathize, to understand how people think and act, and what they need.

You see, the knowledge we often need to solve problems can't be found in books or Google search results, because it exists in people's heads. Whether the person sitting next to you is your client, a teammate, or a user, it's very likely that they have the information you need. So don't be afraid of your ignorance, or of others noticing it. By embracing your ignorance, you can truly begin the process of learning.

Get Past Fear with Users and Design Teams

Julia Choi

Actionable research insights are "design gold" and require working through participant and peer emotions. Fear is a particularly strong human emotion and can put roadblocks between you and others. Here are some tips on preventing fear from derailing effective research.

First, you should know what fear is, when it's a problem, and why it matters in UX research. Fear can help reinforce learning (touching a hot stove, for example), but an unhealthy state of fear can short-circuit uncovered research insights. It can also distract team stakeholders, especially those overextended with unrealistic timelines.

Here are some ways to help study participants move past fear:

- *Plan well and stay flexible:* Avoid rushing through difficult experiences; these require vulnerability and time to think through. If in a group setting, provide an option for participants to share their input privately (for example, via electronic survey). Adapt the session format, including breaks, to help put them at ease. For example, if a person seems uncomfortable with a camera trained on them, stand where their eyes avoid the lens, and be open to turning the camera off.

- *Listen, advocate, and cast off judgment:* Make study intentions and goals clear at the start of your interaction to help them better understand how to help you. Clarify expectations with your participant by letting them know that you are not assessing or scoring their skills. Emphasize that you need their help to understand how the design team should improve the product or service to work better for them.

- *Create distance and use the third person*: It can sometimes be easier to advocate for others rather than ourselves. If the participant seems stuck using *I* or *my* statements, restate the question using their name. For example, use *How would this help Anna?* rather than *How would this help you?*

- *Acknowledge fear*: Rather than dismissing any fears, acknowledge and walk through them together. For example, "I noticed you paused or hesitated here—could you help me understand more?" followed by silence can enable them to share a deeper insight. However, adhere to healthy boundaries and prevent traumatic experiences (that is, avoid an interrogation).

Consider team stakeholder fears as well:

- *Gauge UX design experience*: If the team is inexperienced with UX, they may have unrealistic expectations (inadequate resources, researchers who do development). Invest in educating the team before performing the study or sharing findings and recommendations. Share the value of investing in UX and how it benefits their product/service—this may require patience.

- *Survey resources and time pressure*: Be aware that timing and resources can discourage or distract teams from listening—and might even provoke less rational responses. Distill findings and convey these using objective language with recommendations, and prioritize across project stages.

- *Guide and inform team decisions*: Build empathy using data stories to overcome their fears (for example, a delayed project timeline):
  - Distill concise potential use scenarios, including when/how users experience fear.
  - Add harms/consequences to help prioritize.
  - Gather photographic evidence or add simple visuals.

With a safe space to understand how to serve people and protect their interests, you can empower others to share deeper insights.

Data Alone Does Not Create Empathy— Storytelling Is Key

Kyle Soucy

A key finding from a research study is actionable only if it's shared in a meaningful way and grabs the attention of your stakeholders. As UX researchers, we're often so focused on conducting great interviews that we forget how important it is to also be a great storyteller and to share our findings in ways that get our teams to take action.

Like It or Not, You Must Get Comfortable with Public Speaking

When I first started in this industry, I didn't realize how much public speaking was involved. I'm not talking about presenting at UX conferences either; I'm referring to just the day-to-day when leading a research study: running kickoff meetings, debriefing observers, conducting data analysis workshops, and presenting findings.

Writing a research report is not an effective way to create empathy and tell the user's story. Reports are boring; they're not memorable or actionable— and that's assuming they're even read. Most importantly, data alone does not create empathy. Stories humanize the data by painting a more complete picture so everyone can truly understand what the users are feeling and going through. For instance, I could tell you that 7 out of 10 people couldn't complete a task, but you're not going to truly understand their pain and frustration unless I give you more details about why they are trying to accomplish this task and what it means to them if they can't.

To tell the user's story properly, we must actually get in front of stakeholders and present the research findings. This is your chance to share the user's story and get stakeholders to feel something in a way that reading a report never could.

What's the Secret to Being a Great Presenter and Storyteller?

It's not just the material; it's the ability to connect to your audience and convey emotion. The best speakers actually care about what they're speaking about, and they get you to care too. Whenever you're presenting, it's absolutely key to share how excited and passionate you are. I promise, your energy is contagious, and it's the main difference between a presenter that puts their audience to sleep and one that keeps them on the edge of their seat.

There is one question that you are trying to answer for each audience member when you're storytelling: *Why should I care?* The tricky part is that each person may have a different reason for why they care. It is on us to know what motivates each person at the table and understand their perspective. We must connect with each stakeholder and explain why this should matter to them.

A Word of Caution

Stay away from the dreaded "slideument" (a presentation with too much text that resembles a document). A report and a presentation are two very different deliverables. You should never take your report and slap it into some slides and call it a presentation. If you do this, you're essentially saying, "Here's the report you wouldn't read, I'm going to read it to you." Remember, we get in front of stakeholders and present for a reason. We're trying to get them to connect, to feel, and to take action. You must craft your presentation very differently from a report to keep your audience engaged.

When researchers get welcomed into a person's home or place of work to learn more about their world, it is an honor and a privilege. We owe it to them to tell their story accurately and in a way that gets through to the decision makers. The best researchers are not just great moderators; they're also amazing presenters and storytellers.

Personas with Emotions and Behaviors Are More Valuable

Cindy Brummer

Effective user personas are like having actual contact with real people. They show us the emotional depth of the end users and how they will interact with our products in a variety of contexts.

However, many personas don't take the users' emotions, behaviors, or environments into account and end up becoming a surface description of a stereotype. This can quickly cause misalignment or mismatched expectations between user and designer.

Avoid Weak Personas

Let's start with an example of a less effective user persona. Imagine working on a website for a plumbing company. The designer talks to current or potential plumbing customers. Then they create a persona describing a 45-year-old woman who has a job as a medical assistant and shuttles her kids to after-school activities. Voilà! A demographic-based user persona!

But this is a customer profile, and it gives us only a *surface* understanding of a user. It's also a description of a *good* day. It doesn't include their frustrations, their goals, or even the mental states of when they would call for assistance from a plumber or how they would find the company.

A strong persona would be based on data collected from users drawn directly from our research and observations of real people. Personas also share the users' behaviors, thoughts, motivations, and goals with the rest of our team.

Step 1: Start with Data

Let's strengthen this persona:

- *Collect rich information:* In your interviews, collect emotions, behaviors, and context as well as demographics.
- *Always use real data from real people:* The only fictional part should be name and image. In this case, we'll call her "Renae."
- *Review your interviews:* It's tempting to lean on what you remember from your research. Instead, take the step to externalize all of the data from the interviews so you can see it. You could use software, a spreadsheet, or a whiteboard. Write down the collected data in one place to categorize the information. Be sure not to include any information for Renae that didn't come from real users.
- *Synthesize what matters:* Focus on how users will interact with the plumbing website. For example, we don't need to include that Renae enjoys romance novels or going to the movies.

Now that we have a decent picture of Renae, let's take this profile to the next level.

Step 2: Provide Context

To make the profile into a strong persona, we need to provide relevant context. Chances are strong that Renae will visit our website in a variety of different mental states. She might feel excited about a new home renovation project and call our plumber. She also might feel upset and angry because a water pipe just blew and she needs help.

Review the data from users to think through the many different contexts in which Renae may need a plumber. You can surface these through a variety of methods, including structured brainstorming, mind mapping, or bodystorming. These techniques can create a narrative script through which you can understand what Renae is going through at any given moment.

It's not possible to capture every single emotion of our personas, but by identifying two to three primary ones, you can develop stories that communicate how Renae will use the website, depending on her situation.

For our products to be successful, our personas should consider the emotional, behavioral, and environmental contexts so we can know what's going on inside the minds of the people we're designing for. It's the only way to break free of stereotypes. Otherwise, real users may not identify with the products we build for them.

Educate Your Product Team for Successful User Research

Rachel Young

Your moderator guide is reviewed and the prototype is built, but does your product team understand what to expect during a research session and the role you want them to play in it? What is obvious to an experienced UX practitioner may not be obvious to others. With education and support, you can help make the research process more transparent to your product team.

Educate on What Research Can and Cannot Answer

Your product team will comprise members from product management, development, marketing, and training organizations, each with their own questions about user needs. As you formulate your research plan, involve your team so everyone is aligned on what research can and cannot answer.

- Gather questions from all team members to ensure you've included various perspectives.
- Map your teams' questions to the research methods best suited to answering them. For example, when testing a medical device, the following methods might be appropriate:
  - *User interview*: to gather users' perceptions of the device's usefulness, color, and content
  - *Usability study*: to test comprehension of the instructions and mechanism for loading medication into the device
  - *Tree test*: to measure the intuitiveness of the device's navigation menus

- Provide an overview to your team, explaining the different methods, how they address the teams' key questions, and any gaps where other methods should be used.

Explain What Research Is and What It Is Not

For some team members, research sessions are their first opportunity to hear users' feedback firsthand. Others may have awareness of focus groups, user surveys, or user acceptance testing and expect UX research to be similar.

As a moderator, I'm often asked why I answer users' questions with a question or do not offer help when users struggle. To someone less familiar with UX research, our methods can seem evasive or insensitive. While we understand why we do what we do, it is our job to make that clear to our product team. Take the time to cover a few of the basics:

- A user research session is *not* a demo. Participants need to interact with design concepts in context.
- We conduct research in a one-on-one setting and test with multiple participants to look for emerging trends.
- Participants typically do not receive training before interacting with design concepts.
- The intent of usability research is to uncover where users experience difficulty. It is okay to have users struggle because that tells us where we need to make improvements.

Encourage Observation with Specific Guidelines

Having team members observe research helps demystify the research process and foster shared understanding of end users and their needs. Set up guidelines for your team's participation to ensure sessions run smoothly and provide the most insight.

- Invite team members to observe research sessions and attend debriefs as their schedules permit.
- With each session invite, include sufficient context for observers:
  — Participant schedule
  — Participant demographics
  — Interview guide

— Remote meeting links

— Observer guidelines

- Remind observers *not* to interrupt research sessions; instead, enable a way for their questions to still be heard. Allow team members to send questions electronically to a designated "liaison" who collects and funnels them to the moderator to be asked during wrap-up.

- Provide observers with pointers on how to observe, what to look for, and the difference between an observation and an interpretation.

- Invite team members to take notes, providing a simple notes page to do so. Sections for notes can include the following prompts:

— User

— Task

— Quotes

— Key insights

— Open questions

Your product team is your supporting cast as you plan and execute user research. Taking time to set the stage for them ensures active participation and a shared understanding that will contribute to the overall success of your research.

Design Isn't Just About the Happy Path

Drew Lepp

Optimism has the power to change the world, but an overly idealistic mindset can lead to poor decision-making, sometimes with disastrous results.

In 2012, I was a designer at an education start-up with a mission to teach financial literacy to students in the poorest and least resourced communities. At the time, there was tremendous optimism about how new technology such as iPads, Chromebooks, and digital education applications could radically level the playing field for students.

This positivity energized and pushed our team to create a product we believed would have a real impact for students. Unfortunately, it also created blind spots. There were a few places in which the worst-case scenario—excluding student access to educational materials—became a reality.

Had we asked ourselves "What's the worst that can happen?" earlier, we could have prepared contingencies to prevent us from excluding the exact people we aimed to help.

Why Does This Matter?

Design isn't just about the happy path; it's about how to address possible issues from all angles. What are people more likely to remember: an effortless, easy-to-use experience or an unsatisfying, possibly damaging one?

By identifying both positive and negative outcomes early, you will be able to better design a product experience that is more adaptable to the wideranging realities of the human experience.

Examples of Worst-Case Scenarios

Physical harm and endangerment are often the first things that come to mind when considering negative outcomes. In the 1940s, when the field of human factors was coming of age, the worst-case scenario was obvious: if a cockpit wasn't designed to properly fit the pilot, the pilot could crash—it was a matter of life and death.

With digital products, the worst-case scenarios, with some exceptions, are less defined and more nuanced. In fact, many of the design issues we face have never been seen before in history, and we're all figuring out how to address them together.

While I've found that accessibility and usability issues are generally well understood, topics like emotional health, digital literacy, bias, exclusion, attention, and injustice, while seen as important and worthy causes, typically don't receive the same level of visibility and attention throughout the design process.

Practical Advice

- *Research and co-create:* Work with people who represent a diverse population and/or are subject matter experts. Including a wide range of perspectives will uncover a more complete set of worst-case scenarios.

- *Ask tough questions:* When starting a new project, think through the worst-case scenarios early. For example, who could this harm? Who are we excluding? How does this affect vulnerable populations or those going through hardships? Where can bias occur?

- *Start early:* When starting a new project, think through the worst-case scenarios early and incorporate them into your goals, anti-goals, and design principles.

- *Find a balance:* While it's important to understand risks, it's best to be neither too optimistic nor too pessimistic. Overoptimism can cause blind spots, while focusing too much on where things can go wrong can cause anxiety and distractions and hinder creative solutions.

- *Educate yourself:* Nobody has all the answers. It takes ongoing training, education, research, and introspection to understand how to better design for a broad set of situations.

- *Speak up:* If you encounter something that could be problematic, say something early. Follow up and be persistent if things don't change.

Whenever you start a project, always ask yourself, "What is the worst that can happen?" Though it's not easy, asking about the worst-case scenario early on will lead to better products that can accommodate and respond to a larger range of possible experiences.

Deliver Successful Products Through Common Success Metrics

Martina Borkowsky

The principle that you can improve only what you measure has led to countless ways to measure success: product managers want to see a high conversion rate or adoption, software development wants to deliver a piece of software free of functional bugs, and marketing measures the cost to acquire a new customer. Have you noticed the pattern? All of these key performance indicators (KPIs) are in their own silo—their own little piece that contributes to the whole. By focusing too much on these silos, people tend to lose the big picture.

When people think of UX metrics, they often think of "number of clicks." Numbers are powerful, but if your only goal is to reduce this number, you miss a point. Maybe you added a confirmation dialog to protect customers from unintentionally losing work. This certainly improves the overall experience but also increases the number of clicks—numbers do not always tell you the full story.

Many businesses rely on standard scales such as the System Usability Scale or the Net Promoter Score to evaluate their products. These give you one global number that is supposed to represent user satisfaction. But your value proposition is not objective or comparable to everyone else in your industry. You might produce an app that offers the cheapest flowers online or the most reliable desired-date delivery, or you might offer the most exotic plants—but you're not all of these. You have your own specific goals to differentiate your product in the market.

Focus on the user and collaborate with stakeholders to create meaningful metrics that uncover how you can increase user success and add business value. Ask why teams measure certain KPIs and help anchor them around customer value. You'll quickly realize that your respective goals behind the numbers have more in common than you think.

As a designer you work hard to optimize workflows, and developers ensure they write high-performing, stable code. Your common goal is that you all want to release a product that increases customer efficiency and satisfaction. Different people contribute to this goal through their unique skills.

These collaborative sessions unite teams behind a handful of common success goals unique to your product. Now you can determine how you can best evaluate your individual contribution to these goals with metrics more meaningful than a standardized number:

- Create a customer feedback survey based on unique product success goals.
- Ask questions in a way that provides actionable feedback:
  — For a goal of better storytelling with data, ask users to rate their confidence levels when making decisions. The rating shows you how meaningful (or not) the shown data is. You can allow free-form answers to understand what the user might be missing.
  — For the goal of easier onboarding or increased learnability, ask users to rate if the workflow supports these tasks. If this gets a poor rating, then the objective that you had in mind was not met, and you know it needs more attention.

Collaborating with teammates to determine meaningful KPIs empowers you with quantitative and meaningful user feedback that can uncover the true story of how your users feel about your product. It can show where you didn't deliver as expected and where there are opportunities to improve. It's best to combine surveys with a user interview to add qualitative color. If you use this approach, you will get insight into what truly matters to your customers and what your next steps should be.

Bring Rapid User Research Methods to Agile Teams

Bob Thomas

When I started my 12-year career at a Fortune 100 company as Director of UX Research, my goals included increasing the visibility of UX research and establishing a research strategy. Philosophically, that strategy was simply understanding our customers' needs. Pragmatically, I would describe it as "test early and often."

Traditional UX Research Methods

We did plenty of qualitative usability research using "think-aloud" protocols. We recruited participants and moderated sessions using task-based scenarios, such as: "Imagine you bought a new car and want to insure it. Please show me what you would do." Participants told us what worked for them and where their pain points were. Typically, these studies took 45–60 minutes and involved 10–12 participants. We watched and listened to users—but our stakeholders never did.

It took a total of five weeks for these traditional usability studies:

- *Two weeks:* Recruiting participants and preparing for usability sessions
- *One week:* Running usability sessions
- *Two weeks:* Preparing a readout for our stakeholders, who didn't watch any sessions

We knew five weeks was not going to work in Agile.

Lean UX Research Methods

Beginning in 2015, our UX research team incorporated new usability methods to better meet the needs of our Agile teams. We focused on the actual experience under design, not the final deliverables.

Conflicts do exist between UX and Agile methodologies, including that Agile has a developer focus and UX has a customer focus. Agile relies on short iterations and rapid feedback from customers. As most Agile sprints are two weeks, this leaves little time for discovery research. However, our new Lean UX research methodologies worked in Agile teams after we completed our discovery research and ran traditional usability tests.

In traditional usability testing, our final deliverables were in the form of what Steve Krug of *Don't Make Me Think, Revisited* (New Riders) fame calls the "big honking report." But compiling such reports took time, which was not in Agile's best interests. And you only compile reports for stakeholders who are too busy to watch and listen to their customers.

Our goal was to collaborate. We wanted to help Agile teams test design ideas rapidly, validate/invalidate them with real users, and share UX insights earlier in the process. In our new Lean methodology, we began measuring time in days instead of weeks:

- *One to two days:* Recruiting participants from research panels, based on a 5-to-10-question screener, or recruiting participants from previous studies.

- *One to two days:* Running the usability sessions with 5–10 participants. Stakeholders from Agile teams observed sessions in real time in another observation room, with a UX research team member facilitating. During these observation sessions, stakeholders wrote single observations on physical sticky notes (we now do it remotely with online whiteboards like Miro or Mural).

- *One day:* Running a "Find the Problems" collaboration session with the UX and Agile teams who observed the usability tests. In this session, we:

  — took sticky note observations and grouped them into categories

  — labeled each of those categories

  — voted for which categories were the most important

  — identified the highest-priority categories/problems

- *One day*: Running a follow-up "Find the Solutions" collaboration session that identified and worked out fixes for the problems found in the usability sessions.

No matter the methodology, our UX strategy was still the same. We wanted to test early and often to understand whether we were meeting our customers' needs. An organization that values the user experience can succeed only when everyone—not just UXers—watches and listens. We're all busy, but if we don't watch and listen to our users, then we're relegating ourselves to building user experiences based on subjective opinion.

Scale Research Through Stakeholder Advocacy

Matt DiGirolamo

Whether you're starting as a sole practitioner or entering a somewhat established company, understanding how to scale UX research is critical for advancing your team's UX maturity.

This work doesn't happen all at once. And while it's possible to get from zero buy-in to something greater than that, you should acknowledge that some executive willingness is going to be required to move all of this forward. But how do we get to that point?

Identify Current UX Maturity

Talk to people within the organization and figure out what they do now, what they know about UX, and their sentiment toward it. Map your UX maturity model by identifying gaps between where you want to be, what's missing from the practice, and what is currently happening. There are myriad maturity models; research to find the right one for your practice.

Foster Awareness

Present to stakeholders and provide details about what you do, why you do it, how it's done, and what value you provide. This "road show" can help identify allies in your endeavor. Keep these people close since they provide valuable outreach throughout the organization.

Next, do a quick and scrappy usability study on an area of opportunity. Once you have results to share, tell a compelling story of the work you just did. If value is there and stakeholders are looking to get more, demand for UX work will increase.

Set Up Research Frameworks

Well-defined processes make your UX team more efficient and effective as demand increases. Document your UX research process and templatize everything. Then share the content in an organizationally accessible place. For example:

- Pick a research plan template and use it as a research request intake form.
- Use a method decision framework (like the one created by Christian Rohrer (*https://oreil.ly/KFPbh*)).
- Set up standard ways to recruit participants.

Push Maturity Forward with Team Ops and Quantification

Evolve how UX is perceived and operates by capitalizing on the momentum of increased demand. If value, demand, and executive willingness are there, increased head count will follow! Prepare appropriately:

- Create leveled interview rubrics and scenarios/exercises; determine an equitable hiring process.
- Move from a central consultancy model toward embedding research into product groups. You'll get more in-depth exploration of problems, and researchers will quickly become domain experts with their stakeholder partners.
- Create an internal central library of easily accessible results. Remember: there are many ways to build a functioning research repository, but it is no small task.

Start tracking metrics to inform whether design decisions make a difference:

- Pick a measure that makes sense for your product or practice. Some to consider include SUS, SUPRQ, UMUX, SEQ, and SUMI (but not NPS!).
- Educate your stakeholders on its pros, cons, and proper use.
- Benchmark the current product and track changes to the metric regularly.

Research Democratization

"Research for all" can seem scary in terms of methodological and practical purity, but it can work:

- Leverage your process and templates to have others validate their work.
- Use unmoderated techniques and technology when possible.
- Build out a recruitment panel of your users for quick access.

Your researchers won't be entirely hands-off, though; in its infancy, this program should have checkpoints where researchers collaborate on any work.

Like UX itself, this is an iterative process—flexibility is critical, and all solutions won't work in every case. What doesn't change is the connection with stakeholders, sharing information widely, and constant collaboration. We can be agents of change by following these steps, but without top-down support and resources, we'll struggle to reach a high UX maturity. However, we can tip the scales in our favor by showing, telling, connecting, and producing.

Know When and How to Build a Usability Lab

Rich Buttiglieri

There are several research methods that can make use of a usability lab, such as design walk-throughs, focus groups, and usability studies. A formal usability lab is a useful tool in conducting these studies, though it is not at all mandatory. You can easily run a usability study with just yourself facilitating each participant, either remotely with a screen sharing application or in person.

When Would I Need a Lab?

There are two main functions of a lab:

- Allows observers to watch a session live without risk of interference
- Records the session from different camera angles for later analysis

Watching in person from an observation room encourages team building and collaboration. This is especially important for in-the-moment analysis. You can record most digital interfaces via screen sharing applications such as Zoom, but a lab allows you the ability to record interactions with physical devices. A lab also helps to support the importance of usability research in your organization and promotes customer loyalty by showing them you value their feedback.

How Expensive Is It to Build?

Lab equipment and room modification costs can vary wildly from a few hundred dollars to tens or even hundreds of thousands of dollars. Usability labs come in all shapes and sizes, from remote testing with a screen sharing application to a multiroom suite with remote controllable cameras and a broadcast studio all-in-one computer with switchable sources and the ability to record and broadcast to another room or stream online.

If you are considering building a dedicated lab space, you first need to decide how often the lab will get used and the availability of walled offices or conference rooms you can take over. Also consider all the camera angles needed for remote observers to see what the participant is doing. If you are exclusively testing software on a computer, then a single camera on the participant's face for an inset picture may suffice. However, if you are testing physical products, consider adding more cameras to capture different angles, especially if participants move around a lot.

Physical Space Considerations

Once you have justified the expense of equipment and space, you should consider the following points when planning for a lab:

- Locate a permanent space near the main entrance, on the ground floor if possible, and use good signage so participants and observers can easily find their way.

- Minimize the chances participants may encounter observers—keep the test room and observation room doors separated, including the walking paths.

- Make the test room configurable for all types of studies and participants. Lighting control helps with vision-impaired participants, and movable furniture helps with wheelchair access.

- When the test room shares a wall with the observation room, add soundproofing, such as acoustic wall panels or full deck walls, if possible.

- Make your observation space comfortable and accessible as well, with good chairs, dimmable lighting, good HVAC control, coffee, water, snacks, and maybe even some quiet fidget toys to keep people focused.

- Mount the test room cameras so they are not so conspicuous to participants. They know they are being recorded, but some may get self-conscious if a camera is directly in their line of sight.

- Decide whether a two-way mirror (a.k.a. one-way glass) between the test room and observation room is needed. Observers are able to empathize more when they can see the whole person and not just a video feed with a narrow field of view of the participant. But glass does allow sound to leak, more so than a solid wall, and it can make participants uncomfortable seeing a large "interrogation-like" mirror and knowing they're being watched.

Talk to Customer Support to See What's Tripping Up Users

Dave Connis

UXers should treat any coworkers in frequent communication with customers as deities. Just by doing their jobs, these folks gain tons of insight into the problems users face when using your product. Whatever your company calls them—customer success, customer support (maybe both?)—treat them as a resource for gathering data and challenging hypotheses.

Renaming a Feature

For example, I was a UX writer at an application development platform where customers repeatedly asked for a feature even though it was already in the product. With some research, my team found the current name wasn't connecting with users because it didn't give them any context into what the feature actually did. So it sat untouched in a settings section.

My team was debating a new name when my manager said, "I wonder what users call this specific functionality when they talk to support?" So we walked out of the meeting room and over to a customer support rep and asked them. Sure enough, when customers requested the feature from customer support, most used a specific term that wasn't reflected in the name or in our docs. So instead of coming up with a brand-new name, we just named it what our users already called it.

We checked in with support a month later to see whether customers still asked for that feature, and we found requests had decreased by 68%. From then on, we made checking in with support part of our taxonomy definition process, specifically when renaming or improving text in an existing feature.

Practical Application

Start a relationship with support. Getting access to their customer-centric knowledge is a huge opportunity for getting insights into where customers experience the most friction. The best part is that it just takes a quick email or Slack message. I've never had any support folks tell me to "get lost" when I've reached out to chat, since better design makes their jobs easier.

Here are some things to consider when reaching out:

1. Make it clear that you want to help make their jobs easier by reducing customer friction.

2. Focus on your shared goal: you both want customers to have a better experience.

3. Have a specific question in mind for what you want to learn.

   a. A more specific question could be "How many calls do you get about feature X?" or "Where do you see users get the most confused when doing X?"

   b. More general questions could be "What are some of the things you find yourself saying over and over again when it comes to using the product?" or "What feature is the biggest source of your support tickets?"

Another way to get data from customer support is to get permission to comb through support tickets. If your company has one, visit its online product community and write down what problems continually make users ask for help. If your company has an active social media presence used for customer support, you can gather data from that space too. Once you have a good list, analyze your data, group them by feature, and see where patterns emerge!

Be Prepared When Practicing Ethnography

Meena Kothandaraman

Ethnography is the rich description of a holistic context through observation that often focuses on people, their environment, interactions, artifacts, and overall objectives. Ethnographic practice permits the study of a "bigger picture" to gain perspective and clarify learning objectives. Deep observation uncovers nuances that might otherwise be overlooked and should always be paired with a closing interview to ensure that documented data carries no embedded assumptions.

Ethnography is most effective when conducted in person. Founded in the social sciences, ethnographic practice has become a primary method to study the user experience in-context. The open-ended nature of observation requires considerable preparation, beginning with a clear introduction of the study and the learning objectives, and sharing of the engagement arc. The engagement arc defines the research team's intended interactions with participants. To prepare:

- *Communicate the learning objective and interest in conducting ethnography:* Convey the session intent to the participant with full transparency and understand the rules of their environment (connect directly with your participant and do not communicate "through" someone). Document special requirements demanded by the environment you are visiting (for example, no shoes in the house, steel-toed boots on the warehouse floor, and so on).

- *Define the session boundaries:* Describe your intended engagement arc, the observation duration, your level of interaction (and whether interruptions are allowed, or whether questions are tabled for the closing interview), and finally how the data collected will be used. Listen for and adjust to concerns in the arc where possible. Transparency is imperative to establishing trust.

- *Adjust to the participant's convenience:* Participants are offering their time; be prepared to join them when their tasks occur. Send a meeting invite documenting session details.

- *Share a final confirmation:* Remind participants the day before the session and reconfirm the schedule and engagement arc. Ensure the research team and observers review meeting logistics and break and mini-debrief timings. Establish clear expectations for observer conduct.

- *Assemble your researcher kit, including:*
  — Guiding paperwork (informed consent, closing interview discussion guide and activities)
  — Notepads, a clipboard, and writing implements (analog/digital)
  — Recording equipment and extra batteries (audio/photography/videography, if permitted)
  — A separate timepiece (if phone use is not permitted)
  — Quick, quiet, odor-free nourishment

Participants are now fully informed, and the sessions are scheduled. The ethnographic engagement arc is typically three straightforward steps:

1. *Begin with a tour:* Walk through the area within which the participant exists and elicit details related to the learning objective. Draw a quick map (for future reference), noting people, locations, and areas of interest. Document observations from multiple perspectives.

2. *Study their task(s):* Document the participant space and tasks. Note what they are doing, what tools and resources they are leveraging, and who is involved. Describe every action and every "verb." Employ a "who, what, where, when, why, and how" strategy for your observations. Note that initial observations will be rich and detailed. As time progresses, codify repetitive tasks as a way to streamline observations. Accompany the participant as they move. If you have permission to interrupt the participant, ask short clarifying questions, and table larger inquiries for the closing interview.

3. *Conclude the observation:* Thank the participant for their time, and set the expectation for the closing interview to mitigate any remaining observation-related questions. Recreate the context of the questions you ask in the closing interview to remind the participant. Have the

participant review photographs taken for data transparency and delete any deemed sensitive.

Ethnographic practice results in contextually rich and thick data and benefits from intense and transparent preparation. Employing this method results in a heightened perspective of the people we study and their environments.

Always Do a Test of Your Test

Jacqueline Ouifak

You never know what will happen when you combine a clickable prototype, test script, and test plan with a human being, which is why it is so important to be prepared for the unexpected.

The main goal of conducting user research with a prototype is to learn what is and is not working with your proposed design so that you can make informed decisions for the next design iteration. The last thing you want is the distraction of a broken or misaligned prototype, test script, or test plan that could have been avoided.

In my first formal role as a UX architect, my boss would insist that we make the time to do a dry run for any planned usability testing. I would shudder; it was hard enough to understand the business domain, the usage scenarios, and how the users might use our enterprise app, let alone run a usability test. And a dry run on top of that? I already felt the pressure of thinking through all the possible user paths, building them out in the clickable prototype, and having to write a cohesive test plan.

The day came for me to finally do usability testing on my own. I was so confident in my ability that I thought I'd be just fine doing the dry run the afternoon before the first actual test day. As I went through the dry run, one thing after another went wrong, and slowly my heart sank to my toes as I heard my boss's voice in my head saying, "This is why we do this." The test script didn't match the paths the dry-run participant was taking; the scenario details didn't match the participant's real-world scenarios; and the test took too long to get through.

I spent that afternoon frantically fixing things, and luckily the live test went smoothly the next day. I can't imagine what a disaster it would have been without that dry run. Fifteen years later, I give the same words of advice to my team, especially since prototyping tools are more powerful, interaction behavior is more subtle, and usability test goals are more ambitious. This makes the art of coordination and preparation even trickier.

No matter how big or small the user research session, take the time to do a dry run with anyone who can fill in as a representative user:

- *Ensure that the prototype is running on a platform that a participant can access and that it is displaying correctly*: Does the design work on the other end?

- *Read your script out loud*: Does it make sense to the person hearing it? Did you miss anything? Does the scenario and task order make sense?

- *Look for the unexpected*: Did the dry-run participant take any unintended paths? Is it important enough to adjust the prototype, or can you talk through it instead?

- *Ensure everything works together*: Are there problems that you could have avoided with your setup, your synchronization with the prototype, script, and/or plan?

Give yourself at least a half day to respond to what comes up, so that when the real sessions happen, you can focus on what you need to get out of them and not regret that you weren't better prepared.

You can never know what will happen when you combine a human being with your clickable prototype, test script, and test plan, but you can at least be better prepared for what might happen by doing a dry-run test. Never let them see you sweat!

Observed Behavior Is the Gold Standard

Kaaren Hanson

One of the biggest errors designers (and teams) make is listening to customers—that's right, *listening* to customers.

What am I saying? Is it that customers don't matter (Nope! Customers matter a lot.) Is it that people aren't trustworthy? (I do believe that most people are trustworthy). The problem is...

There is often a gap—and sometimes even a chasm—between what people say and what they do. The cleverest people and teams primarily focus on what people do.

Let's look at some examples.

As discussed in Part V, *Don't Ask Users to Predict the Future*, page 214, people are not good at predicting their future behavior. Think back to your last New Year's resolution. Did you achieve it? If so, you're one of the few—people are notoriously hopeful. For example, millions of people optimistically join gyms each December. Yet just a month or so later, most stop going. The podcast Planet Money reported (*https://oreil.ly/gu81v*) on a gym that had 6,000 total members, even though the facility could hold only 300 people, because most never show. What we want to do and what we actually do are not the same. As UX professionals, we need to account for this gap when conducting research.

People don't just fail at predicting their future behavior; they are also inaccurate at reporting their past behavior. For example, many years ago, Intuit acquired a payroll company with a less-than-ideal experience. We did some behavioral benchmarking and deliberately compared what people said and what they actually did. In the benchmarking, a large sample of customers did a series of payroll tasks (printing a paycheck, adding a new employee). Immediately after completing each task, we asked the customer whether they had been successful, and we also noted if they had actually done the activity correctly. In some cases, what people said was true (85% said they added direct deposit correctly, and 85% actually had). In other cases, they were very far off (75% thought they had correctly printed the paycheck, but only half actually did the task correctly; 80% thought they had correctly added a new employee, but fewer than half had actually succeeded). Fortunately, the team jumped into a (badly needed) redesign and brought the actual success rates up to world-class levels (90%+). If the team had simply relied on self-reported success, they could have easily passed over fixing a very poor experience.

What people say can be seductive. Self-report data is easy to obtain, and we want to believe people are accurate. At Wells Fargo, I frequently ran into data that was labeled "success," when in reality it was only self-reported success. Again, we saw gaps of up to 30% between what people said and what they did. The best designers and teams take what people say with a grain of salt. Moreover, they build habits that keep behaviors top-of-mind. For example, they focus on actual success rates as a key outcome. This might be from tracking instrumented code or from simply observing people doing activities "in the wild." As teams design research studies, note-taking sections will deliberately be labeled "what people say" and "what people do." Creating deliberate and explicit reinforcement of customer behaviors will help our teams stop relying on what people say.

The trickiest part is that sometimes what people say and what they do actually aligns. Unfortunately, we cannot predict when people are accurate and when they aren't. Self-report data is an important source of directional inspiration, but it is not a sound source of rigor. That is why we'll all be better off watching what people do. Observing a person's behavior is truly the gold standard.

Assess Usefulness and Desirability Early in Product Development

Michael Hawley

The protocols and best practices in usability testing are designed specifically to find problems. If your question is "What is getting in the way of someone completing their task?" then usability testing is the right method. The approach is inherently based on observing behavior: learning from what people do, not from attitudes and perceptions. Finding usability problems is the focus, and the method works great.

However, while creating highly usable products is always a goal, the challenge of creating successful products is increasingly complex. Good usability is an expectation. The challenge now is to create products that stand out in the marketplace and offer significant value. We shouldn't focus on usability problems in the early stages of concepting, since we're still determining features and content. Instead, we should focus on product-market fit, the overall usefulness of a design, how well it will reflect the company's brand, and the potential to deliver real value. For example:

- What data is missing that can help customers make an informed decision?

- Does the branding reflect desired attributes?

- Does the overall organization of content align with customer perspectives?

- Which conceptual approach to a design problem will resonate best with customers?

- How would the design fit into customer workflow?

- What level of personalization or gamification would be appropriate for the product?

To get at these questions, we could turn to foundational research before we create any design sketches. But in today's Agile-oriented business culture, it can be challenging to find the time for the in-depth research needed to answer these questions ahead of time. Additionally, it can be difficult for research participants to articulate their perspective without reacting to some concept or artifact.

To give context then, we can show participants a sketch, a prototype, or a design composition and collect feedback. This approach starts to look like usability testing: one-on-one sessions with a moderator and design artifacts. But researching usefulness and desirability is not usability testing—we are not just trying to find usability bugs. Instead, we want to assess potential value and attitudes toward a design. To accomplish this, we can still have participants walk through a design, but we need to modify the approach.

It can be tempting to simply ask research participants, "What do you think of this page?" But absent any frame of reference or prior experience, participants are not likely to offer real critical thinking. It can also be tempting to ask, "Would you use this?" This approach is inherently flawed, as people are notoriously bad at predicting their future behavior.

Assessing usefulness and desirability in a proposed concept requires us to be intentional about putting the participant in the right mindset to offer critical thinking and providing a structure for tangible feedback. Consider the following best practices when creating scripts and protocols to assess usefulness and desirability:

- Before showing designs, explicitly ask participants to discuss current pain points and needs and to tell stories about past challenges to establish a mindset and a reference point for evaluation.

- Remind participants of their stories and expectations during the design walk-through, and utilize those points to prompt honest assessment of the new design.

- Utilize structured feedback protocols such as Product Reaction Cards, Collaging, or Emotion Cards to prompt discussion on the proposed design.

- Favor participant-directed "tasks" and exploration of design concepts.

- Emphasize questions on personalization in summary questions—for example, "If this could be personalized just for you, what would that include?"

There is an art to uncovering customer attitudes and preferences on early design concepts. Be intentional about modifications to traditional usability testing protocols to uncover insights that can guide your team.

Know the Core Elements of Usability Research

Amanda Mattson

Usability studies are one of the fundamental methodologies of UX research. These involve a moderator guiding a participant through a structured series of tasks on an interface and/or a prototype. During your career, you may be asked to conduct usability studies on your own prototypes or on prototypes others have created, or you may need to consume research output—which is why it is key to have a handle on the basics of usability studies and when they are appropriate.

We conduct usability studies to learn whether an interface's functionality (1) aligns with how users expect it to work and (2) supports getting them through a workflow. Conducting these studies early (and often) during the product development life cycle helps a product be more user friendly from the onset, avoiding expensive redesign later. There are three main elements to usability studies:

1. *Tasks:* With the interface in one hand and research/business goals in the other, establish the critical tasks participants will accomplish and document the correct/desired path(s). If you are working with a prototype designed by someone other than you, verify the paths you've written are correct and achievable. Ensure each task is straightforward and has a single goal. If a task has participants doing too many things, it probably needs to be broken into a second task.

2. *Moderation:* Moderation for usability is a little different than it is for user interviews. Here, you lead the participant through your task list and questions, rather than facilitating a more natural conversation like you would in user interviews. The key to moderation is probing on interesting tidbits and asking *why* multiple times to undercover root causes. During usability studies, moderators need to balance staying quiet while observing, asking questions at the right time, and taking notes on key elements.

3. *Participants:* To collect data that accurately informs your design, you need to recruit the right participants. By creating clear and precise screener criteria (see Part V, *The Right Screener Sets Up Your Recruit and Research for Success*, page 199), you or your recruiting agency will be able to successfully target the correct population for your study.

These are the three core elements of usability research. Next, consider the following pointers to have a smooth first usability study:

1. *Prototypes:* Get the prototype ASAP. Researchers working with prototypes often don't get them until the last second, which isn't conducive to effective research. To increase the likelihood of having enough time with the prototype, ensure your design colleagues or clients know you need time to learn the prototype's quirks and correctly document what you expect participants to do.

2. *Pilots:* Run a pilot with a coworker, friend, or family member. This will help determine timing and cadence, as well as uncover any initial friction points with or bugs in the prototype so you can flag them before your actual study day. Pilots help to ensure a successful first study day.

3. *Recruiting:* As soon as you have the screener criteria, start recruitment. To increase the probability of recruiting the right participants, allow at least two weeks between screener completion and the beginning of the study. Proper recruiting is time consuming but worth it. If you're working with a recruitment agency, schedule a project kick-off meeting with them to properly set expectations. A warm handoff is key; simply emailing them to kick off a recruiting effort leads to miscommunication.

Naturally, there is much more to usability studies than this, but once you've mastered these core elements and tips, you will be equipped to run meaningful usability studies that lead to actionable insights for you and your team.

Don't Underestimate the Power of Coworkers as Usability Participants

Daniel Diener

In digital product development, iterative and continuous testing is a success factor. But conducting research with the target group can be time consuming and expensive. Testing products with colleagues from outside the project offers a valuable alternative. They can be easily recruited and will give you relevant feedback on the usability of your product. Because if the product doesn't work for your colleagues, it will fail for your customers as well.

Especially if you have a product for which it is difficult to recruit the target group, usability research with colleagues is essential. When we started to work in sprints in our organization, user research was always kicked out of the sprint planning because recruiting our customers was always time consuming and tedious. We also have only a limited number of total customers, and therefore we need to use their time wisely. This has led to issues conducting iterative usability research with customers quickly, which can be the death knell for fast research.

With our two-week sprints, we wanted to test small iterations continuously. So we started conducting usability studies with colleagues from other departments, whether accounting, purchasing, or another area of the organization. We recruited five colleagues for each sprint and tested our product in small 15-minute sessions. At every session, there were different colleagues joining. We were able to improve our product before we tested it with customers, saving them for higher-stakes formative research. Without these initial studies with colleagues, we would have a significantly lower product quality today.

You can achieve success through continuity. These three tips will help you achieve success in testing with colleagues:

1. *Build a database of internal participants so you can recruit them easily and quickly:* By doing this, you can make sure that you have as many contacts as possible who want to test products, and you avoid annoying colleagues who don't want to participate.

2. *Ritualize the research:* Make it as easy as possible for teams and participants to plan the sessions. For example, you can schedule sessions every second Friday. This way, the teams know when their testing object has to be ready, and the study participants are aware they have to take time for the research.

3. *Build up a community event:* We do a big event every quarter called the Usability Testing Marketplace. Market booths are set up at which teams offer their products for testing. Participants can move from stand to stand and do as much product testing as they want. However, it is important that a session should be no longer than 5–10 minutes. The best thing about the Marketplace is that teams can exchange learnings with each other and colleagues can test different products with little effort. The biggest effect is that you build and strengthen your community.

Test your product as often as you can. Usability research with coworkers is of course no substitute for customer feedback. However, it helps to improve your product. Use your colleagues' feedback early so you can reserve the customers' time for initial formative research and feedback on higher-fidelity prototypes.

Include Nonusers in Your User Research

Becca Kennedy

For product research, the easiest way to recruit participants is often by contacting existing users. Although it's great to interact with customers, many design teams stop here and miss out on learning from nonusers too. Maybe your team needs fresh ideas, or your company is venturing into new markets. Maybe your current users are too change-averse to give unbiased feedback.

A balanced UX research strategy often includes *nonusers*—people who do not currently use your product but potentially could.

Nonusers Can Help Uncover Gaps

Current users give helpful feedback in some contexts, but there is a lot to learn from nonusers. Maybe they are former users—why did they stop? Maybe they use a competitor—what do they like and dislike about that one?

When I conducted UX research with a financial services company, we interviewed nonusers who were interested in personal finance but didn't use this company's offerings. We learned that, unrelated to specific product features or education, sometimes people don't like traditional investing because it conflicts with their personal values, such as by profiting from industries that harm the environment or society. If we had interviewed only current customers, we'd have missed out on understanding why some people avoid investing.

Nonusers Can Supplement User Research

Some nonusers don't use a product because they aren't experts in a specialized field. Why do we care about these folks? Occasionally, it can be difficult or expensive to access specialized audiences. Research with nonusers can supplement the research you are doing with expert users.

I once worked with a software company whose users managed indoor farms. Most research was with existing customers, like getting feedback on new features or visiting facilities to learn how the software fit into workflows. But the specialized participant pool was small, and we didn't want to bother the same customers repeatedly. We supplemented research with *nonexpert* nonusers to get feedback, for example, on the public-facing website, while saving customers' feedback for the in-the-weeds stuff. (No pun intended!)

Nonusers can provide insight on design aspects that don't require subject matter expertise, if it is safe to do so. It might be quicker and easier to use a general participant pool, rather than experts, to test tasks such as signing up for an account or finding general information within a product, like profile details.

How to Include Nonusers

There are a lot of benefits to researching with nonusers, but it can be tricky. Here are some ways to set yourself up for success:

- Consider the research goals and identify when research requires
  - current users (to understand current user flows, for example)
  - a more general audience of nonusers (to evaluate your product's learnability, for instance)
  - a combination of both (you might, for example, aim to recruit 7 users and 3 nonusers for 10 interviews)
- Recruiting nonusers can be hard. Work with a recruiting agency or get creative with finding participants online (as on social media) or in person.
- Ask potential participants to complete a screener about relevant behaviors, motivations, tools, and so on. This will help you identify who fits the profile of a user, whether they already use the product, or whether they use something else.
- If the product requires specialized knowledge, give nonusers instructional information up front, like the context of use, who the users typically are, and any jargon that they might need to know.

Strategic user research often requires recruiting outside your product's customers to help you uncover knowledge gaps or supplement research with current users. It takes extra effort but will result in a more robust and balanced understanding of your product's user experience.

Plan User Research with the Customer Question Board

Julia Cowing

In the age of businesses focusing on their customers, anchoring product road maps around the customer is critical. But often there are so many customer-reported problems that teams can become overwhelmed. A way to help teams plan and prioritize user research is through a framework called the Customer Question Board.

Here are five benefits of the Customer Question Board:

1. Focuses conversations on customer problems
2. Builds a prioritized, stakeholder-aligned research backlog
3. Sets up user insights to be more effective and impactful
4. Helps teams map customer questions to research methods
5. Encourages a user research mindset and culture

The Question Board is simply that: it's a board where a team places all of its questions related to the customer and their experiences with your design.

Here's how to use this framework:

1. *Prepare:* The first step is to make sure the members of your interdisciplinary team understand the objective of the exercise. Send out a calendar invite preparing them to think about customer questions they might have. For example, a question might be, "How easy is it for users to use the feature we're building?"
2. *Prompt:* During the session, ask the team members to place their questions on sticky notes. Virtual whiteboards also work!

3. *Discuss and cluster:* Have one team member explain their question and place their sticky note in a central area. Then have everyone with a similar question place their sticky note next to that first one. By doing this real-time affinity mapping, teammates will be encouraged to discuss the mutual data they need to understand the customer.

4. *Attitude versus behavior:* Create an x-axis labeled *attitude* to *behavior*, left to right. Next, ask the team if each cluster of questions is about what the customer is *saying* or about what the customer is actually *doing*. Place "saying" questions on the left; these are attitudinal questions. Place "doing" questions on the right; these are observational, behavioral questions.

5. *Qual versus quant:* Create a y-axis labeled *qualitative* to *quantitative*, top to bottom. You should now have a 2x2 matrix. Ask the team if the question clusters are about figuring out *why* something might be happening. These are qualitative questions, searching for insights; put these in the top left or right quadrant, depending on their horizontal position in Step 4. Or if the question clusters are about *how often* something is happening, these are quantitative questions and go on the bottom left or right quadrant, depending on their horizontal position in Step 4.

6. *Vote:* There should now be clusters of questions in four quadrants. Have the team vote on which cluster is most important to explore first.

7. *Map questions to methods:* Each quadrant of the matrix has different user research methods to gather needed data:

    - *Discovering insights (top left):* Customer interviews and diary studies are methods for these attitudinal questions.

    - *Observing usage (top right):* Usability research and eye tracking are methods for these observational and behavioral questions.

    - *Validating insights (bottom left):* Card sorting and surveys are methods to quantify and validate customer attitudes.

    - *Validating usage (bottom right):* Product analytics and A/B testing are methods for these questions to quantify and validate usage problem areas.

Moving questions around on a wall helps teams immerse themselves in the problem space.

The Customer Question Board is part of the user researcher's toolkit but can be used by stakeholders to help generate ideas for design exploration. It helps a team practice making proactive steps toward learning, because when work is already vetted and grounded by customer research, product stakeholders have more confidence in the broader scope of work.

If Designing Survey Questions Were Easy, There'd Be No Garbage Data

Annie Persson

Surveys are a fast, efficient, and popular way to complement and validate your qualitative data. They're one of the most accessible quantitative methodologies you can employ, but there's more to a survey than just quickly writing questions. How you design questions can greatly impact the validity of your research.

I designed a survey to understand how my employer used our company knowledge base and community forum. I thought I knew what I was doing by designing the question closed-ended with a "type in your answer" response option. But then I got a whiff of my own garbage data.

> My question (garbage in): Do you refer instructors or students to the knowledge base and community forum for getting help with their answers?
>
> My responses (garbage out): Yes, No, Sometimes

Part of designing questions for a survey is deciding whether to use open-ended questions (respondents answer in their own words) or closed-ended questions (respondents select from a set of choices). If your goal is to achieve statistical significance, capture data from a large target population using closed-ended questions.

If I could tell my younger self how to rewrite that question today, I would construct it as closed-ended, group it into separate questions, and eliminate the leading verbiage:

> Do you refer instructors to the Knowledge Base? Yes | No
>
> Do you refer instructors to the Community Forum? Yes | No

With this question design, I can compare the frequency with which the company refers instructors to the knowledge base versus the community forum. Depending on the response, I could branch the question to follow up with *Why?* and capture the motivation and attitude behind the behavior.

Here are some best practices of what to do and what to avoid when designing your survey questions.

Creating Questions

Do ask one thing at a time. Group questions of the same topic together and proceed from general to specific for questions on the same topic. It's good to ask closed-ended questions to capture quantitative data and to follow up with open-ended questions to understand the *why*—which requires coding answers to find themes and quantify results. Consider open-ended questions when you aren't sure of the range of responses, and order your questions with easy questions first and difficult/sensitive questions at the end. Finally, position questions that address topics in the introduction section of the survey at the very beginning.

Avoid *and*—it can turn one question into two. Stay clear of leading or loaded questions. Don't use single or double negatives. Avoid questions that rely on memory of past events or ranges of time or that predict the future.

Wording Questions

Use simple and familiar words. Apply simple syntax, and exercise specific and concrete terms as opposed to abstract and general wording.

Avoid generalized terms and absolutes (*always, never*). Stay clear of jargon, technical words, and slang. Don't use words with ambiguous meaning; you want all respondents to interpret questions the same way.

Creating Response Options

Use branched-based responses to skip and display relevant questions. A 5-to-7-point Likert scale is steady and reliable. Include *Not applicable* or *Don't use* as answer options to prevent respondents from skipping questions.

Avoid saying *I don't know*—research shows people select it for ambivalence and self-protection. Check for overlapping responses (for example, *5–10 min.* and *10–15 min.*).

We all strive for clean data. The quality of your survey responses is only as good as the design of your questions. To ensure you're on the right track, always pilot test your surveys and iterate on them until they're ready for launch. Even the best of us need to be told to take out the trash!

The Right Screener Sets Up Your Recruit and Research for Success

Katelyn Thompson

A *screener* is a set of questions used to select participants for a research activity. Good research starts with talking to the "right" audience, and it's important to craft a precise screener to find the best participants.

Define and Outline Your Criteria

Start by defining and then prioritizing the demographics *and* behaviors you want for participants based on your research goals. Are you looking to talk to Android users? People who travel internationally? Prioritizing your criteria will help if you need to make compromises once recruiting begins. Use specific measures instead of vague terms that could be interpreted differently by respondents.

Next, consider the situation for your research and what you'll be asking of your participants. Do they need to be willing to share their location with the app? Use open-ended questions to gauge a respondent's ability to articulate their thoughts, which is essential for interviews. Finally, determine any quotas or mixes of criteria you want. Do you want half of your respondents to be current customers? Do you want a mix of ages or locations?

Now you have an outline for your screening criteria ready for review by your team:

- Uses an Android phone (must have)
- Traveled internationally within past six months
- App use
  - — 50% use our app at least once a month
  - — 50% don't use our app

- Age

  — 25% are 18–25 years old

  — *etc.*

Determine the Method for Asking Your Questions

Common methods for fielding a screener include online surveys, phone calls, or video responses. Sometimes your research tool locks you into a recruiting method. Many remote testing tools require automatic screening, where you do not have the option to review responses and manually admit or disqualify respondents. There is no "best" method—you'll need to decide what's right for you based on the amount of time and money you have. See the next chapter for more information about working with recruiters.

Write the Screener Questions

With your screening criteria and method defined, it's time to start writing your screener questions. You will want to use language familiar to respondents—avoid acronyms or jargon if speaking to the general population. Always give an "out" to your screener questions—include a "None of the above" or "Prefer not to answer" option. The same principles that apply to good survey design apply to screeners.

Ideally, your most restrictive screener questions should come toward the beginning of the screener. However, you want to avoid "giving away" the answers to questions. For example, when recruiting for a certain smartphone, starting broad and narrowing down is a good approach—start with "Which of the following devices do you use regularly?" and then ask "What type of smartphone do you use?" However, if your criteria are very specialized, it may be helpful to start off with pointed questions.

Ensure your question formatting is appropriate for your method. A long list of options is easy to read on a screen, but it's not ideal while listening over the phone. Consider splitting out "select all" questions into multiple "yes/no" questions for phone screens.

As with good design, it's important to iterate on your screener questions. During research you may notice gaps in your criteria to change for the next recruit. To save time for future research activities, consider creating a standard set of questions for your product or business as a starting point.

Select Your Participants

If you aren't using an automated screening tool, the final step will be selecting your participants. Confirm that everyone matches your initial criteria and that you've filled quotas. You're now ready to conduct your research with the right participants who meet your target criteria!

Know Best Practices for Working with a Recruiter

Ellen Finn

When starting a new UX project, it is important to determine whether you will need a dedicated participant recruiter. In selecting a recruiter, you should always confirm that your recruiter comes equipped with their own database. Most participant recruiters have a very diverse population of people (age, gender, ethnicity, geographic location, occupation, and so on) in a specific file system or database, which they can easily draw on to fill various types of studies. Additionally, recruiters will often utilize other outlets, such as social media, online recruiting forums, job advertising sites, and word of mouth, to get their post distributed to the appropriate audience.

Timeline Management and Recruiter/Client Communication

Time management and communication are critical aspects of interacting with a recruiter. Most recruiting projects are set under tight deadlines, so the recruiter will need to set a schedule that works in conjunction with the client. Communication is key to the recruiter and their client working in unison. When the screener is finalized and sent over to the recruiter to begin, the recruiter will need to create a schedule (Excel) and posting (Word) to present to their database. After the posting is distributed and responses are collected, screening phone calls can begin. Usually the recruiter will have a few questions for the client about screener discrepancies or inconsistencies, as they may arise during a screening phone call. A timely response to the recruiter's questions is essential to the flow of the recruit.

Once a few participants are scheduled for a given study, the recruiter will send the client a spreadsheet with the participants' demographics and screener answers. Updates will continue to be sent every few days until the study is filled. It is important that the client gives feedback as early as possible about participants who may not exactly match the criteria, as they may need to be replaced.

Participant Confirmation

Once the recruit is finalized, it is usually the responsibility of the recruiter to send the confirmation emails to the panel. This letter typically includes the appointment time, location, driving directions, consent forms, and any necessary login links. The recruiter and client usually collaborate on the content of the letter and send it out about five days before each appointment. To ensure all questions are answered and everyone will show up for their appointment, the recruiter will place a reminder call or send a text to each candidate the day before their session.

On-Call Duties

Your recruiter should expect to be on-call for the duration of the study days. They are the intermediary between the client and the participant. Both parties will reach out to the recruiter if there are issues on the day of the study. Some complications that tend to come up may include participants running late or Zoom links not functioning properly. The recruiter is there to ensure that the logistics run smoothly for all parties before, during, and after sessions.

Study Aftermath

Once the project is complete, it is essential for the client to give feedback to the recruiter. A wrap-up phone call or email is an appropriate end point to most UX projects. Both good and bad comments are welcome and will be added to that person's profile in the database. The recruiter will usually follow up with the participants as well, to ensure that they have received their incentives and were satisfied with the process. Happy participants and clients keep a recruiter in business and a database growing.

You Don't Need a Lot of Money to Recruit Participants

Thomas Yung

As a UX researcher, finding and recruiting the right participants for a study is essential. If that sounds scary, here's the good news: you don't need a lot of money to recruit participants, if you are willing to put in the time. There are several ways to source participants. The first—using an *external agency*—is expensive. The cheaper way is to *find them yourself*, but this will take a lot more of your time. Here are some strategies that you can use.

Representative Users and Screeners

Representative users are the target audience for your study. To increase the validity of your research, use *personas* and *user behaviors* to create your screener instead of just demographics.

Sample Size

Next, you must decide how many participants to recruit. Essentially, you're looking to have as many users as possible until you reach insights saturation, when you no longer gain new insights from having more users test your system. For most *qualitative* studies, five to eight users per user group will provide enough insight to derive useful design decisions. However, for *quantitative* methods (such as surveys), you will need to have a much larger sample size.

Incentives and Compensation

- This depends on the target audience. Find out what they normally make per hour. Compensate them based on their expertise and required time investment. Typical compensation in 2021 for a member of the general public ranges from $75 to $125, depending on where you live.
- For surveys, give out gift cards or enter them in a raffle for prizes. For moderated studies, raffles are not recommended because not everyone will be compensated for their time (only the winners of the raffle).
- It's not necessary to compensate internal employees, but small tokens of appreciation (parking or cafeteria vouchers, for example) will help.

Posting Ads and Screener

Here are some tips for recruiting participants on your own using ads and a screener:

- Go to where your target audience is likely to be:
  — Internal employees: have departmental managers reach out to their employees
  — Online: Facebook and/or LinkedIn groups, Twitter posts with targeted hashtags, Reddit, Slack
  — Physical locations: coffee shops, grocery stores
- Put a link to the screener in the ad.
- The screener should weed out anyone who would potentially bias the study.
- Pick participants from your screener results. Follow up with each one with an actual phone call to make sure they are not just doing it for the compensation.

Build a Panel Yourself

I recommend building your own research panel using an opt-in form and database. It might take months to compile such a list, but it will save you time in the long run. Building your own panel makes it easier to conduct rolling research in which you loop in regular research sprints with Agile teams. You can then select from your panel of willing participants a week in advance of each sprint. Inexpensive ways to build a panel of participants include the following:

- Ask current participants if it's okay to contact them again for future studies or have them refer others.
- Create a dedicated website and opt-in form to collect info from participants who are willing to help in studies at a moment's notice.
- Post targeted social media ads leading to your dedicated website.

Final Thoughts

There are two types of costs associated with recruitment: having the right number of representative users, and using the right ways to find them. You'll save money by doing much of the legwork yourself; when done strategically, you'll save time as well.

You Need Good Planning for a Diary Study

Mac Hasley

Diary studies take a lot of time, energy, and resources. They also give us a lot of findings, access, and intimate knowledge about someone's day-to-day life. There's a lot to consider when pulling together study logistics. Here's some advice on how to deal with all those *a lots*.

Align Research Questions with a Diary's Structure

Structure your study to take advantage of the "in the moment" capacity diaries offer.

Have participants complete a log entry or answer a few quick questions every time they complete a certain interaction. This immediacy is key to unveiling what's new, interesting, and nuanced about an action. Often facets of experiences feel major in the moment but are forgotten upon reflection.

As you write a research question or hypothesis, decide on a corresponding trigger. That is, give participants a specific prompt that's informative but not leading—that lets them know when to capture or document.

Set a Timeframe that Gathers Insights Before Participants Lose Interest

Even your most engaged participants will get sick of logging six months in. Ask yourself: "How long do I *really* need to run this study to get valuable findings?" And also: "How much data will I *really* have the time to tag, sort, and dig through?"

Choose the Right Tool for Your Goals, Budget, and Timeline

Now comes the critical moment at which you have to make a choice: how do you get participants' thoughts and feelings into your organized research folder? Here are a few options:

- *Go vintage:* Send your participants a physical diary. Have them write in it. Get them to send it back. Avoid this unless you have a very small budget and a long timeline; you'll get all your results in around the same time, making it hard to sort and prioritize as you go, and you'll have to read potentially awful handwriting and manually import your data.

- *DIY it:* Use a method of communication your audience is already familiar with—email, WhatsApp, SMS, Facebook, and so on. Depending on the platform you choose and the digital tools at your disposal, you might be able to collect multimedia entries or automatically send reminders. This also offers you some level of search capacity, but exporting and sorting data will still be a bit of a pain.

- *Use a research tool:* Tools like dscout and Indeemo offer feature sets that make running a successful diary study *a lot* easier. A smart diary study tool will allow participants to log multimedia entries. It'll work well on mobile, send your recruits reminders, and help you recruit participants. One of the biggest advantages of using a research tool is the capacity to easily deal with your qualitative data.

Be Strict in Your Recruit

Aim for fewer but better participants. Consider the conclusions you want to make and then be honest about the sample needed. Whatever your needs are, focus on capturing a few *very solid* recruits from each bucket who together weave a complete picture.

Decide How Participants Will Log

There are a few different techniques you can employ for data collection:

- *Interval-contingent protocol:* You pick regular, predetermined reporting intervals. This is most useful if the behavior you're trying to observe isn't situation dependent.

- *In-situ logging:* When a specific event occurs, your participant logs.

- *Signal-contingent protocol*: Some sort of set alert or alarm tells participants when to complete an entry.

Though diary studies take a lot of time and energy, they afford you a lot of intimate knowledge about someone's day-to-day life. They're a source of truly rich insight that can change the way you think about problems, products, and the people who will someday use those products.

Improve Usability Testing with Task Cards

Todd Zazelenchuk

Sometimes the smallest detail can make a process work better. Remove a pebble from your shoe to make a long hike more comfortable. Add a drop of oil to a bicycle chain to smooth out your ride. In our field of UX design and research, incorporating task cards into our usability studies provides a small but powerful boost to improve study execution and quality.

Task cards are a simple concept: they are individual cards that display the printed task for the study participant to read. They are printed on paper for in-person studies or presented on-screen for online usability sessions. Asking participants to read tasks out loud, rather than the researcher reading them, has several advantages:

- Task cards empower participants during a usability study. Participants must contend with many variables when researchers deliver tasks verbally, including the speaker's volume level, enunciation, pace of delivery, and occasional accent. In addition, the participant must deal with their own short-term memory limitations in remembering task details. When researchers use task cards in their studies, it is not uncommon to see participants reference them multiple times while performing a task, particularly when tasks are complex.

- After participants have read the task (preferably aloud), the researcher can ask them to paraphrase the written task to confirm their understanding before beginning. This can help engage the participants, course correct their interpretation if needed, save time, and ultimately improve the reliability of study results. It also helps participants avoid trying to complete tasks when they have misunderstood or misheard the researchers' instructions. This reduces wasted study time when participants follow wrong paths because of a misunderstanding.

- Task cards help eliminate any unconscious bias or inconsistency between sessions that can happen when a researcher verbally delivers tasks. This is especially important over the course of a long study, where fatigue can become a factor for researchers.

To realize the full benefits of utilizing task cards in your next usability study:

- Begin by writing high-quality tasks that are clear and relevant to participants. This effort is critical regardless of how you deliver your tasks and is a prerequisite to planning and executing any successful usability study.

- Select a legible font and appropriate font size to ensure readability. Given that a primary goal of task cards is to communicate your tasks more effectively, you'll want to be confident that participants can read them without issue. Arial, Helvetica, and Open Sans are all effective, reliable fonts for both printed cards and online displays.

- Ensure the task card is always accessible to your participants as they work to complete the task. For printed task cards, simply hand the card to the participant at the beginning of each task and let them reference it as they go. For digital task cards, display them on a dedicated slide, PDF, or web page that the participant can access as needed while working with the design being tested.

- Include a visible card ID number somewhere on the card to help quickly identify and organize tasks at any point during your study. Making the ID number subtle yet visible (for example, grey text in a consistent location) will make this item "fade into the background" of the card until needed.

With well-written tasks in place, augmented by this new presentation method, you are ready to raise your research game to the next level. Whether your next study is in-person or a moderated remote format, using task cards—just like the drop of oil on your bike chain—will make it run even more smoothly.

Apply the Butterfly Approach to Interviews and Testing

Stephen Denning

As a researcher, you've probably been in situations in which you are conducting an interview, you've built a good rapport with the participant, and the conversation is flowing—but the things they are saying, though relevant, are not lining up with the interview script that you've so carefully constructed. Or maybe you're running a usability study, and the participant deviates during one task and begins completing a task further down your list. How should you respond to these situations? Should you pull the participant back into your line of thinking or roll with it?

In these situations, it is useful to consider the *butterfly approach*. When you watch a butterfly collecting nectar, it doesn't move in a systematic way from flower to adjacent flower. More often it moves in a seemingly random pattern between flowers, stopping for varying lengths of time. The reason for this is that butterflies see things in a different way than we do, detecting nectar by using their ability to see ultraviolet light to distinguish between different flowers. In other words, their perception of the task is unique to them.

In the same way, when it comes to user research, you should be prepared to flow in a way that may not appear systematic to you but instead fits with the way that the user perceives the topic. This flexibility allows you to see the situation from their perspective, rather than risking getting hemmed in by your own knowledge or assumptions.

For this to work well, however, there are a few things to consider:

1. *Be clear in your goals:* It is crucial that you are very clear about what you need to get out of a research session. Whether it is understanding how a participant approaches something in their life or work, how they feel about an experience, how a product compares with their expectations, or

how easily they can complete a task, having a clear goal or goals for the research allows you to more easily focus on what is important during the session and to clearly articulate questions and tasks to participants.

2. *Establish priorities:* Once you have created questions or tasks to help you achieve your goals, determine which are must-have aspects to focus on and which are your nice-to-have aspects. This will help you make a judgment call during the session about whether you can spend time on a deviation, or whether you need to bring the participant back to the topic that moves you toward your goal.

3. *Build flexibility into your approach:* For you to be able to move between topics or tasks on your priority list in a flexible way, you need to build flexibility into the way you manage the session. For example:

   - Where possible, word questions or tasks in a way that doesn't imply a fixed order.

   - Take notes in a way that allows you to jump between the topic areas easily. For instance, have separate sheets for each topic area, or take notes on a tablet using a digital note-taking app that links your handwriting with an audio recording of the session.

   - Make a recording of each session, so that if you are caught off guard by a change in direction and miss something that is said or done, you can refer back to the recording after the session is over.

By having that structure of goals, priorities, and a flexible approach in place, you let the participant guide you more than your scripts or plans. This ability to be guided by the participant, without losing control of your goals or the session, puts you in the best place to understand their perspective. This is, after all, the key purpose of user research. Don't just read the questions...have a conversation!

Don't Ask Users to Predict the Future

Ingrid Cruz

When conducting user research, the goal is to identify our most pressing questions and find the shortest path to reliable answers. As such, it can be tempting to ask users direct and hypothetical questions—like how they think they might behave in a made-up scenario. When we do this, we ask the user to predict what they might do in the future. This may seem convenient, but it's problematic because any number of unforeseeable circumstances or external factors could influence what they would actually do. Most people tend to recall fairly accurately what they did in the past, but when we ask them about what they might do in the future—even if they predict those things with a high level of confidence—they're usually wildly inaccurate.

Hypothetical Scenarios Produce Unreliable Feedback

For example, if I ask you what you will eat for breakfast tomorrow, you will probably give me the best-case scenario. You might tell me that you will wake up early and make a hearty, organic, well-balanced breakfast. You will eat in peace as you contemplate the day and probably simultaneously meditate and do your morning workout.

Reality might look very different. Maybe you realize you're not in the mood for a healthy breakfast, and instead you grab a day-old donut on your way out the door and wash it down with a venti-quad-eight-pump-vanilla-latte with whipped cream and extra caramel drizzle.

Asking someone about the future will give you the glossy image of their very best behavior. That's why we should never take such predictions at face value or make design decisions based on them. Instead, ask questions about the past. What did they have for breakfast yesterday, and why? What about the day before? You'll soon see some patterns and understand what this person's morning routine and breakfast habits look like most days.

The Importance of a Good Research Question

Before we do any research, we should always define our main research question in writing. This is for internal use and articulates what we are looking to learn. For example: *How can we better understand the breakfast habits of working professionals between the ages of 30 and 40?*

Research questions are different from the questions we ask users. We would not ask a user: *What are your breakfast habits?* If we did, we would get the oatmeal-and-yoga story all over again. A better question would be: *Tell me about what you did yesterday from the time you woke up until you arrived at work.* This question would help us understand what they ate, when and where they ate it, and how that fits into their morning routine.

By zooming out with our questions, we get to capture things at the periphery that help us understand motivations and thought processes. If our questions are too granular and usability-based, we miss important factors that could influence the experience. Instead of starting questions with *What, How*, and *When*, prompt them with *Tell me about the last time you....* They will likely answer this question in the form of a story, which is much more helpful than a simple answer.

Asking questions this way can feel counterintuitive. As human beings, we typically want to get right to the bottom of things, and we ask questions accordingly. That's why I always encourage designers to develop a research plan in writing prior to any research project. Define one or two high-level, big research questions; then decide on the best method for answering them. If you do user interviews, design your questions in advance. And most importantly, don't ask the user to predict what they might do tomorrow, because they don't know.

Ask Participants to Tell You What You Don't Know to Ask

Amanda Rotondo

You've done stakeholder research, read your background materials, and composed your masterfully symphonic interview guide. You're ready to excavate the entirety of your topic area, gather the knowledge necessary to trudge across the wasteland of the unknown, and forge a miracle for your desperate and painfully grateful end user.

Except that, in the last 15 seconds of your research, your participant drops a bomb as delicate as a silken-robed piranha, and you are left standing cold and alone in a frozen tundra, faced with what you didn't know that you didn't know.

In medicine, it's called the *doorknob question*. It's when, halfway out the door, a patient asks their doctor a question that adds a layer of complexity and urgency to their until-now mundane conversation: "Oh, should I worry that I've been fainting several times a day?"

What???

The doctor's question here might be "How could you not have told me this?" Meanwhile, the patient's response is "How could you not have asked?" For us, the researcher is the doctor, and the participant is the patient.

Most often, your background work and thoughtful questions will have hedged against any surprise revelations. But what a researcher stands to lose is greater than the job of adding one easy question to the end of each session:

"Is there anything I haven't asked you that you think is important for me to know?"

This one little question flips the power dynamic and shocks the participant into taking momentary ownership of the conversation. So many times I've asked that question and gotten "Oh! Ummm..."—and in that moment of unadjudicated cognition, they let forth the truth of their situation.

I discovered the power of this question when investigating an internal tool used to aggregate product details across disparate ecommerce sites. The data entry team's inputs were consistently invalid and unusable, costing the company a staggering amount of time and money. I prepped thoroughly for user interviews as part of a redesign project, yet still went into the interviews having absolutely no indication of what could be happening.

After three user interviews, I still had no clue where things were going awry. Half out of frustration, I asked, "Is there anything I haven't asked you that you think is important for me to know?" The user stumbled for a second and then said, "Well, I know a lot of people aren't worrying about correct data because of the volume bonus."

What???

It turns out the floor managers' bonus structure prioritized the volume of data their teams input over the quality. Some managers were giving a monetary bonus to the individual on their team who input the most data, hoping that added motivation would result in a bigger bonus for themselves. This was not official; upper management did not know about it. And yet it was leading to sloppy inputs and upending the entire system.

This would have gone completely under the radar if I hadn't asked the question that empowered the interviewee to think beyond my direct questions and tell me what I didn't know to ask.

Asking this question has given me insights, driven me to seek clarifications, led me to add or alter questions to strengthen subsequent interviews, and significantly increased the value of my resulting reports. Additionally, adding it to a script is extremely low cost. It takes seconds to ask, and the worst that can happen is the interviewee says, "Uh, no—you covered it." Meanwhile, the best that can happen is you discover a key insight that enlightens your work and leads to huge benefits for your stakeholders and your users.

Leverage Your "Psychologist Voice" for Effective UX Research Moderation

Dan Berlin

Mindfulness is the act of paying full attention to the here and now—something humans are traditionally not very good at. Our minds tend to wander away from the present, especially if we are bored, troubled, or on "autopilot." When moderating UX research, we should be fully present. We need to listen to the participant's every word and guide the conversation toward actionable insights. Just as a musician focuses on the music, artists on their subjects, and drivers on the road ahead, researchers need something on which to focus to keep them in the moment.

That something is our voice. We can be much more present in the conversation if we are mindful of not only *what* we're saying but also *how* we're saying it. Making the conscious decision to use a soothing voice and keep an even pace while talking will both put participants at ease and keep you focused on the present. Even after more than a decade of employing this technique, it hasn't become rote. To this day, it helps keep me focused and present. It's in-session meditation. Additionally, using this soothing voice—instead of your everyday conversational voice—should remind you to channel your inner psychologist to probe for root causes of your participants' answers. Indeed, I call this approach "Using your Psychologist Voice."

To use your Psychologist Voice, keep these five components of your voice in mind when chatting with participants:

- *Pace:* Keep an even, calm pace; don't rush
- *Volume:* Perfectly audible, but not too loud
- *Timbre:* Warm and soothing; soften your voice a bit

- *Emphasis and pause:* Stress important words and stop to emphasize key points
- *Inflection:* Not monotone, but stay within a given range—don't let your voice get too high or low

Once you have found your Psychologist Voice, leverage it to be mindful during research sessions and to act like a psychologist:

- *Get at root causes by probing:* A psychologist is always probing! Ask *why* multiple times in different ways to get at root causes of participants' wants and needs. There are plenty of ways to ask why without using the word *why*:
  - "Tell me more about that."
  - "What makes you say that?"
  - "How does that make you feel?"
- *Have participants answer their own questions:* A classic psychologist maneuver is to turn patients' questions back onto them. See what participants think the answers might be before you answer their questions (if you do at all). For example, if the participant asks, "Is that what it's supposed to do?" you respond with, "What would you expect it to do?" Remember, however, that you shouldn't do this for *every* question, or else the participant may get frustrated with your constant question dodging.
- *Practice quiet patience to get more data:* Psychologists are often quiet to allow their patients to speak, and they're always calm, even if their patients are frustrating. The same holds for researchers: stay generally quiet to let participants fill quiet spaces, and never get outwardly frustrated, as it could make the participant uncomfortable.

When people meditate, they learn to bring their wandering mind back by continually focusing on their breathing. This gets easier over time and can eventually become a part of one's daily life. The same holds true of using your Psychologist Voice: being conscious of the vocal techniques you choose keeps you in the moment and will be a gentle reminder that you should be the "Experience Psychologist" who gets at the root causes of participants' unmet needs.

Tell the User's Story via Effective Research Reports

Susan Mercer

The goal of the research report is twofold: it should describe what you learned and what you recommend doing next. For usability studies, it should also clearly document what was tested, as that can help document why the team made certain design decisions.

To influence design decisions, you must communicate your findings clearly. Speak directly to your business stakeholders, product managers, designers, data analysts, and engineers. Avoid UX jargon, use clear business language, and speak to their interests. You never know where a report will land later, so explain the details as if the reader has no idea what the study was about.

Pick a format that your organization will most likely read and reference. A slide deck is the norm for many companies; others use wiki pages or formal written documents.

Organize your report to tell a story of your research. The following sections work quite well.

Background

Provide the background that led to the research. This might be a finding from your site analytics that you want to better understand, or a big question that the team wants to investigate.

Goals

Concisely outline the goals for the research study in three or four bullet points. This provides context to your audience to help them understand the study approach.

Executive Summary

Provide a short, scannable bullet-point summary. The first bullet point should outline the methodology. The remaining points should highlight key findings and recommendations. Use bold words to emphasize key concepts.

Methodology

Provide a short section outlining the chosen method and how you implemented it. Include an anonymous description of the participants and their relevant characteristics so your audience can be confident that you included the right people.

Detailed Findings

- Focus on observable things—what you *saw* participants doing and what they *said* aloud.
- Make your inferences and interpretations of the data clear. This will ensure your team understands the data supporting your interpretation.
- Report counts of participants who did/said certain things; avoid percentages or some/many/most categorizations.
  - *Poor:* Most participants liked having large product images.
  - *Good:* Seven of ten participants clicked to view one or more large product images.
- Include verbatim participant quotes or short video clips. Hearing directly from the participant is quite impactful.
- If you are reporting on a usability study, include screenshots of exactly what was tested. In a slide-deck report, add callouts to clearly identify the design elements that relate to the findings.
- Include both positive findings and areas for improvement. This makes it clear that you're being objective. Product teams more easily accept critical findings if you praise what works well. It communicates what not to change.

Summary

Bring it up a level and summarize your findings. Your summary should outline your key themes. Consider adding indented bullet points referring back to your data.

Recommendations

Here's the "So what?" part. Tell the team what they should do next based on what you learned.

- *Point out the issues, not the solutions:* Tell the team *what* should be done, but give the product teams room to determine *how* to do it. For example, "make the button more noticeable" is more collaborative than "make the button red."

- *Make your recommendations actionable:* Recommend the team do specific things that are possible within the known constraints. Or ask them to consider adding a project to the road map to fix an issue if it can't be addressed immediately.

- *Tie your recommendations back to business goals:* Your stakeholders must balance business goals and user needs, which are not always aligned. If you show that your recommendations can help accomplish their business goals, the recommendations are much more likely to be implemented.

Contributors

Aaron Parker

Aaron Parker is a UX and product specialist based in the UK. He has a background in psychology and human-computer interaction and completed a PhD for which he looked into how people communicate through technology. He has spent most of his career working at digital agencies, where he's led UX, research, and product strategy for some of the UK's best-loved brands. Most recently, Aaron has been building a new design and product team for the software development company Amdaris.

Design for Users, Not Usability Studies, page 102

Adam Connor

Adam Connor's work focuses on helping teams and organizations strengthen and grow their human-centered design and innovation capabilities. As a design leader, Adam's work blends systems thinking, HCD, anthropology, and organizational behavior to foster more collaborative, creative, and customer-centric organizations. He has coached and trained teams across the world and from industry-leading organizations such as Google, Disney, Fidelity, and Twitter. In 2015 he and coauthor Aaron Irizarry released *Discussing Design: Improving Communication & Collaboration Through Critique* (O'Reilly). His thoughts on collaboration and design can be found at *adamconnor.com* and *discussingdesign.com*.

Remember the Four Questions of Critique, page 70

Al Lopez

Al Lopez is a user experience designer based in the St. Louis area. Their love of technology started as a kid in the '90s when their uncle gave them a computer and told them to have at it. They have been on a quest to understand how humans interact with technology ever since. Their passions are spreading technology to underserved communities, increasing accessibility in everything they build, and pizza. In their spare time, they teach user experience at LaunchCode's CoderGirl and hang with their German Shepherd, Jupiter. Al is also the copresident of the Saint Louis User Experience group. They hold monthly meetings to teach everything user experience, user research, and customer experience related.

Work Together to Create Inclusive Products, page 46

Amanda Mattson

Amanda Mattson can be found conducting user interviews and usability studies and adding emojis to Slack as an experience researcher at Mad*Pow. She was born in Santiago, Chile, was adopted when she was three months old, and has lived in Massachusetts ever since. She is currently surviving the pandemic in Cambridge with her fiancé, Jack, their two cats, Oliver and Babylady, their dog, Potato, and about 25 house plants. Amanda holds her MS from Bentley University in Human Factors in Information Design. Before that, she got her BA in Psychology and fine arts from Hartwick College. She is passionate about helping people with her work and always enjoys lending a hand to people who are new to the UX field. In her free time, Amanda loves listening to murder podcasts, researching new recipes, raising carnivorous plants, and eating gluten-free baked goods.

Know the Core Elements of Usability Research, page 187

Amanda Rotondo

Amanda Rotondo focuses her work on one core belief: technology-based experiences should empower people and lift them up. She has applied this belief across UX design, research, and strategy, both as a consultant and in-house, for over 20 years. In her current role as VP of experience strategy at Digitas, Amanda sets the vision for digital experiences, aligning clients' goals with the needs of users to resolve experience gaps and provide value for all. She has also had a robust academic career, earning an MA in media effects from the Pennsylvania State University and a PhD in human-computer interaction from Rensselaer Polytechnic Institute, where her research focused on sociotechnical systems. Her dissertation specifically investigated the role of empathy in online collaborative work. She continues to teach, publish, and present broadly on these topics.

Ask Participants to Tell You What You Don't Know to Ask, page 216

Andrea Mancini

Andrea Mancini is an Italian designer with a deep knowledge of the internet and the web. After earning a high school degree in electronics and graduating in industrial design, he started a career in the web design field. In 2011 he joined Aruba.it, an emergent ISP, cloud, and trust services provider. At Aruba.it Andrea built the user experience department from scratch. Today, after 10 years, Aruba.it is a major company in Italy, and every day Andrea inspires a group of 20 specialists to design the best services on the market. A digital enthusiast since childhood, he is a speaker at conferences and webinars, but with one condition: he participates only if memes are allowed in keynotes. In love with his country, Andrea works tirelessly for a better digitalization of Italy for the Italian people.

Best and Last Impressions Are Lasting Impressions, page 61

Andrew Wirtanen

Andrew Wirtanen is a principal product designer at Citrix in Raleigh, North Carolina, where he designs enterprise software for both end users and administrators. He has worked in UX since 2006, starting his career as a user researcher and designer at a few agencies before joining Citrix. In North Carolina, Andrew has been very involved with his local User Experience Professionals Association chapter, the Triangle UXPA. He has been an event organizer, chapter president (2013), and UX Y'all conference cochair (2019 and 2020) and is a member of the advisory board as of 2021. Andrew holds a master's degree in Human Factors in Information Design from Bentley University in Waltham, Massachusetts. He is @awirtanen on Twitter.

You're Never Done Learning, page 7

Andy Knight

Andy Knight is a digital product designer with a 30-year track record of creating simple customer experiences across a variety of industries. Most recently, he was Head of Digital Design at TIAA. Prior to TIAA, he spent 10 years with Fidelity Investments, where he led the design of Fidelity's industry-leading retirement planning tools. He also cofounded the health-tech start-up meQuilibrium. These days, he works with select clients as principal of Ronin Digital Design.

Always Go for the Why—the Immutable Basis of Great Design, page 138

Ann Chadwick-Dias

Ann Chadwick-Dias is an independent consultant and founder of Synergy User Research, Design, & Accessibility (SURDA), LLC (*http://surdallc.com*). Prior to starting her own consultancy, she worked in various roles ranging from UX researcher and UX designer to UX director/manager and corporate DEI advocate for 25+ years. She has presented on topics related to universal usability and inclusive design at numerous conferences, including Aging by Design, User Experience Professionals Association (UXPA) International, World Usability Day, CSUN's International Technology & Persons with Disabilities, Human-Computer Interaction, ACM's Universal Usability, Human Factors in Computing Systems (CHI), and many more. She has also taught in Bentley University's certificate program for Human Factors in Information Design and periodically guest lectures at various colleges and universities. She has numerous publications in several fields of study across industry and academic settings, including human-computer interaction, inclusive design, psychobiology, and developmental psycholinguistics.

Design for Universal Usability, page 51

Anna Iurchenko

Anna Iurchenko (@anatinge) is a designer and visual thinker. She works at Google Health, where she creates products that apply artificial intelligence to help doctors better diagnose diseases and generally help people live healthier lives. Sketching and visual storytelling is an important part of Anna's design process. She sketches to visualize user experience, convey ideas, and make complex concepts clearer with visuals. She shares her approach for visual collaboration at her workshops and in the blog *Sketch It!* (*https://sketchit.co*) Anna also coleads the San Francisco chapter of the global Interaction Design Association (IxDA), mentors designers, and speaks at conferences across the globe.

Improve Communication and Encourage Collaboration Using Sketches, page 74

Annie Persson

Annie Persson (@ampersson) is a user experience researcher with over 15 years of experience in the UX field operating as a writer, designer, and researcher. She has presented at local and national conferences and is cofounder of Rocky Mountain Active 20-30, a nonprofit organization benefiting at-risk youth. Annie received her second master's degree in Human Factors in Information Design from Bentley University. When she's not obsessing over UX, Annie is thrifting for items to resell on her eBay store, Thriftluxnstuff—an endeavor inspired by an ethnography study she conducted on sustainable fashion. Based in Colorado, Annie is also the proud mom of two puggles, Opie and Opal.

If Designing Survey Questions Were Easy, There'd Be No Garbage Data, page 196

Audrey Bryson

As a recovering industrial designer, **Audrey Bryson** (@brysonsbest) is able to blur the lines between ergonomics and interaction design. She currently works for an audiovisual electronics company with the ambitious goal of making AV environments less intimidating.

Your First Idea Is Sometimes Your Worst Idea, page 110

Becca Kennedy

Becca Kennedy (@becca_kennedy) is a UX researcher and strategist with a PhD in Human Factors Psychology. She is a consultant, writer, speaker, and mentor on topics related to design research and entrepreneurship. She lives in Albany, New York, with her husband, cats, and many shelves of books and records.

Include Nonusers in Your User Research, page 191

Benson Chan

With 15 years of product management and design experience, **Benson Chan** has helped launch from concept to highly scaled products at Amazon and Microsoft. He currently leads a Design and Research team to invent how Amazon's Alexa digital assistant becomes more personalized.

Sell Your Design Ideas with Trust and Insights, page 79

Bob Thomas

Bob Thomas (@bobthomas) runs his own user research consultancy, Bob Thomas, User Research Consultant, LLC. Previously, he was director of user research at Liberty Mutual Insurance, where he worked for 11 years building a user research practice and managing a co-located team of 10 user researchers. His background includes user experience research, usability testing, graphic design, and product management. He has 25 years of experience in the technology industry. Bob has presented at local, national, and international UXPA conferences. He is on the UXPA Boston Board of Directors. He holds an MS in Human Factors in Information Design from Bentley University and an MBA in international business and marketing from Suffolk University.

Bring Rapid User Research Methods to Agile Teams, page 167

Brian Sullivan

Brian Sullivan is a director of user research and design strategy, where he is working to transform the travel industry through design thinking, user research, accessibility, and service design. Brian's global workshops have trained more than one thousand people globally on four continents. Brian cofounded the UX certification program at Southern Methodist University, where he teaches young professionals how to build usable products with research and design thinking. Graduates from this program have gone on to work at Fortune 500 companies. He teaches graduate design students at University of North Texas.

Frame the Opportunity Before Brainstorming the Solution, page 104

Catherine Dubut

Catherine Dubut leads teams to bridge experiences across digital and analog journeys. She manages UX strategy at Samsung Electronics America and has spent her career as a hybrid UX designer and researcher in roles at REI, Intuit, Socrata, and Bolt Peters. Catherine previously led Hexagon UX Seattle and shared her work at UXPA, IA Summit, Women Talk Design, GIANT, and World IA Day. A Seattle resident, Catherine grew up in the Bay Area and studied music performance and cultural anthropology at Stanford University.

Amplify Your Value by Finding Advocates Outside Your Team, page 17

Chris Callaghan

Chris Callaghan (@CallaghanDesign) is a UX and optimization specialist approaching 20 years of industry experience and has established UX teams and practices at a number of agencies and start-ups. He currently leads a UX team at a global network agency, conducting UX strategy, user research, design, testing, UX analytics, and conversion rate optimization. Over the years, Chris has developed his design intuition from hundreds of hours of usability testing live sites, mobile apps, and prototypes in a range of sectors, with a particular focus on ecommerce. Educated in product design and interactive media, Chris is also UX Master Certified by Nielsen Norman Group and a Certified Usability Analyst by Human Factors International.

When Prototyping, Consider Both Visual Fidelity and Functional Fidelity, page 42

Christopher Coy

Christopher Coy is an artist, an entrepreneur, and a design thinking workshop facilitator. He consults organizations of all sizes on the ways that human-centered design principles unlock positive impacts in their communities.

Legacy Product? Imagine You're Restoring an Old Farmhouse, page 98

Christopher S. LaRoche

Christopher S. LaRoche (@silvaire) works as a senior user experience consultant at the Massachusetts Institute of Technology (MIT). His work focuses on researching and evaluating the accessibility and usability of websites and software. His interest also includes promoting inclusive design. He has worked as a technical writer, an information designer, a user researcher, and a UX consultant in his career.

Inclusive Design Creates Products that Work for Everyone, page 53

Christy Ennis-Kloote

Christy Ennis-Kloote is an eternal optimist with a high capacity for details managing an amazing team of 20+ design consultants. Her background in industrial design and human-centered design leads her to understand users' needs and answer them in the execution of the product. Christy channels her passion for sharing with the community into leading teams through MWUX and IxDA Grand Rapids, and as cofounder of Ladies that UX Grand Rapids. She also enjoys giving back by teaching locally through coLearning at The Factory and at Kendall College of Art and Design. In these courses she has covered concepts and methodologies of product and user experience design through various digital formats such as mobile, web, or desktop applications. Christy's range of work is primarily in design leadership, with a focused interest in connected experiences and product design strategy.

Embrace a Shared Cadence to Avoid Silos, page 83

Cindy Brummer

Cindy Brummer (@cindybrummer) is founder and creative director of Standard Beagle Studio, a user experience design consultancy based in Austin. Cindy founded Standard Beagle in 2012, and she currently leads a small but mighty team of UX designers and developers, helping clients improve their services and technology and make more delightful experiences for their customers. Cindy is passionate about UX design and about sharing her experience. She received a master of professional studies degree in UX design from the Maryland Institute College of Art in April 2021, teaches UX design as an adjunct instructor in the McCombs School of Business at the University of

Texas at Austin, and founded the group UX in ATX. For fun, Cindy teaches boxing and is a certified Level 1 boxing coach through USA Boxing.

Personas with Emotions and Behaviors Are More Valuable, page 157

Daniel Diener

Daniel Diener is a UX designer within the marketing communications department at Porsche AG. As part of the UX design operations team, he supports internal product teams in creating digital experiences at Porsche. True to the motto "Looking good is not enough," the UX Design Ops Team wants to design all digital applications optimally for users—customers, dealers, and employees. The central team conveys a uniform understanding and a common user experience approach in order to work as efficiently as possible and to deliver high-quality results. Through a team of experts (designers, researchers, writers, and developers), they develop tools, educate methods, and create a Porsche design culture. Products that are influenced by the UX design operations team include *porsche.com* (*https://porsche.com*), My Porsche, and Porsche Online Shop, as well as the Porsche Car Configurator.

Don't Underestimate the Power of Coworkers as Usability Participants, page 189

Danielle Cooley

Danielle Cooley is founder and principal of DGCooley & Co., a bespoke strategy and research consulting practice. For more than 20 years, her work in research and design has been applied to a variety of digital and analog products for such companies as Hyundai, Pfizer, Graco, Enterprise Rent-a-Car, Fidelity Investments, MasterCard, and more. A frequent conference speaker, Danielle has a BE in Biomedical and Electrical Engineering from Vanderbilt University and an MS in Human Factors in Information Design from Bentley University. You can find Danielle online at the DGCooley & Co. website (*http://dgcooley.com*), and she tweets sporadically *@dgcooley*.

The Participant's Well-Being Is Your Responsibility, page 140.

Darren Hood

Darren Hood (@darrenhood, @uxuncensored, and @TheWorldofUX) is extremely passionate about all things UX, holding 20+ years of experience in human-computer interaction and interaction design, and for the last 16+ years working full-time in the world of UX, CX, and LX. Darren's professional footprint spans such organizations as Ford Motor Company, General Motors, Bosch, Ryder, Cengage Learning, Duracell, Caterpillar, and USA Networks, to name a few. He also serves as adjunct professor for Kent State University's UX Design master's program and as an adjunct at Lawrence Technological University in Southfield, Michigan, and he leads UX workshops at Grand Circus in Detroit, Michigan. Darren is also the host of *The World of UX* podcast. He is currently pursuing a PhD in educational leadership from Northcentral University.

Know the Difference Between Experience Mapping and Journey Mapping, page 27

Dave Connis

Dave Connis has been obsessed with words for over 10 years. During a brief stint as a junior UX developer, he found that he loved the UX part more than the developing part, so he took on a job as a technical writer. From there, he quickly discovered UX writing, and it was like the heavens opened up and doves descended from the sky offering him a path that combined all of his favorite things (UX and words). Now he's a senior documentation and UX writer at Contrast Security, a start-up that helps developers keep vulnerabilities out of their code.

Talk to Customer Support to See What's Tripping Up Users, page 175

Drew Condon

Drew Condon has been a UX practitioner in the Boston area for over fifteen years. His main interests are design strategy, system design, and organizational design. He has worked in a variety of UX management, leadership, and individual contributor design roles across consumer and B2B products within both larger enterprises companies (IBM, HubSpot, Compete) and early stage startups (RunKeeper, Mylestone). He also is cofounder of a company that helps folks in the seafood industry grow their business through better compliance and automation. Drew has a B.A. in Communications from Boston University and a M.B.A. and Masters of Human Factors in Information Design from Bentley University.

Think Synthetically to Design Systematically, page 59

Drew Lepp

Drew Lepp is a product designer focused on creating principled and thoughtful designs at scale, with a special interest in how technology can facilitate real-world opportunities and connections. Over the past five years, she has worked with large Silicon Valley technology companies to bring the voice of users directly into the product experience. Prior to that, she ran a UX design agency in Washington, DC, and designed web and mobile apps for early-stage start-ups, Fortune 500 companies, and a bit of everything in between. In her free time, you can find her creating her own artisanal perfumes or daydreaming of sailing around the South Pacific.

Design Isn't Just About the Happy Path, page 163

Dwayne Hill

Born and raised in Milwaukee, Wisconsin, product designer **Dwayne Hill** has always had a penchant for how experiences make people feel, and for the excitement of business. He has utilized the knowledge he's gained working in industries such as fashion, retail, nonprofit financial counseling, and business development to craft experiences that provide value to the user and a competitive advantage for the business. It's important for Dwayne to build positive relationships and collaborate with others to solve complex problems. Dwayne's favorite food is chocolate chip cookies with walnuts, and yes, cookies are food. He also enjoys spending time with family, playing sports, reading, writing, and traveling. Dwayne will never pass up an opportunity to hang out on a front porch and sip sweet tea or lemonade while elegantly maneuvering a rocking chair. You can blame that on several years spent living in the South.

Understand and Speak the Language of Business, page 13

Eduardo Ortiz

Eduardo Ortiz is chief executive officer, partner, and cofounder at Coforma. He most recently was executive creative director at the US Digital Service. He has over 19 years of experience as a software engineer, UX designer, and information architect. He cofounded Project100, a women-led, women-focused organization that helped elect 100 progressive women candidates to congressional office in the 2018 elections. He is a cofounder of BKUX, one of the fastest-growing design communities in New York, and is a local leader for the Interaction Design Association (IxDA). Before cofounding Coforma, Eduardo was design director at Ralph Lauren, where he oversaw the replatforming effort for its ecommerce properties and led the team that made the platform accessible to all. He cherishes working as an advocate and helping to bring a sense of normalcy to those in need.

Implement Service Design in Your Practice, page 36

Ellen Finn

Ellen Finn is owner and director of UX Recruiting. She holds a bachelor of arts in psychology from the University of Massachusetts and has been working in the recruiting field for 15 years. She collaborates regularly with UX companies in the Boston area and nationwide to locate, screen, and schedule research participants. In her free time, Ellen enjoys reading, exercising, and playing board games. Ellen lives in Massachusetts with her husband and three children.

Know Best Practices for Working with a Recruiter, page 202

Emily Roche

Emily Roche is an indie freelance content strategist based in New York City. She joins cross-functional teams to organize and guide content through site redesigns, migrations, and other digital experiences. In her travels, she's helped clients such as the MTA, Volkswagen, JPMorgan Chase, and others up their content game. Emily got her start as a writer in the dot-com era and brings an editorial mindset to her work. She's thrilled that content strategy brings out the Dr. Librarian and Ms. Editor sides of her professional personality. If you ask her what's really fun about content strategy, she'll tell you that seeing more people join the discipline is pretty rad. It's wonderful to see so many people working their magic to improve how content is presented on sites, interactions, and whatever comes next. She hopes her chapter shows others how they can help make content sparkle.

Break Your Lorem Ipsum Habit: Sketch with Words!, page 135

Erin Malone

Erin Malone is the principal of Experience Matters Design, specializing in designing systems, complex tools, user research, and social interfaces, and is currently working with the Anti-Defamation League's Center for Technology and Society. She was a founding member of the IA Institute, was chief editor of Boxes and Arrows for five years, and is coauthor of the book *Designing Social Interfaces*, 2nd and 1st editions (O'Reilly). She holds a BFA in Communication Design from East Carolina University and an MFA in Graphic Design from Rochester Institute of Technology. She has been working in digital interactive spaces since 1993. She is also Chair of the Interaction Design BFA program at California College of the Arts, where she teaches Visual Interaction Design, among other classes, and challenges her students to understand grids and color as well as all the Gestalt principles.

Follow These Principles of Gestalt for Better UX Designs, page 63

Eva Kaniasty

Eva Kaniasty is the founder of a user experience consultancy providing UX strategy, research, and interaction design since 2010. She received her master's degree in human factors from Bentley University, chairs the User-Centered Design program at Brandeis University's Graduate Professional Studies division, and is the former president of UXPA Boston.

So You Want to Be a UX Consultant, page 9

Frances Close

Frances Close has a passion for craft and getting the details right. She is a firm believer that functionality is just as important as aesthetics, and she bridges the two to create and deliver meaningful experiences. As a design lead, she leads multidisciplinary project teams and works closely with stakeholders to set strategic vision and goals. She enjoys working through complex problems with cross-functional teams. Frances has a background in product design, branding, visual design, and creative direction and has worked in agencies in Chicago and Michigan. She also has a passion for her community and has taught a user experience course at a local co-learning lab, along with volunteering with several design associations such as AIGA and Midwest UX.

User Experience Extends Beyond the Digital Realm, page 25

Gail Giacobbe

Gail Giacobbe lives in Seattle and has worked in tech product innovation for 20 years. She currently serves as general manager, customer success engineering for Microsoft. Her team of designers, content creators, product managers, and data scientists deliver content, knowledge, and expertise to help Microsoft 365 customers succeed. Gail holds a BA in anthropology and religion from Princeton and an MA in teaching from Brown. Gail is an active community volunteer and nonprofit board member, and she's a frequent public speaker on leadership, career development, product management, and gender equity. She and her husband enjoy traveling the globe with their kids, collecting art, and adventuring in the mountains of the Pacific Northwest.

Design Customer Experiences, Not Features, page 30

Georgiy Chernyavsky

Georgiy Chernyavsky is a designer with a passion to create digital products that help companies reach business goals. Throughout his career he has worked in various agencies, until finally finding his passion in complex enterprise digital products. In his work Georgiy enjoys approaching challenges, overcoming obstacles, and bringing people together. He advocates design in organizations, promotes cross-disciplinary collaboration, and helps design teams be more efficient.

Define What Your Design Does Not Do, page 55

Helmut Degen

Helmut Degen is a senior experience architect for software-intensive products and a researcher in human-computer interaction. He performs business, domain, and user research as well as designing and evaluating applications and services. He specializes in applications and services for complex, industrial domains. Helmut's design aims to optimize for value, efficiency, and remembrance. Helmut won a Red Dot Award for one of his designs in 2019. He is cochair of the AI in HCI conference. In his new book *Every Click Counts: Design for Efficiency (https://oreil.ly/tTmxy)*, he describes a comprehensive architecture and design framework with a focus on efficiency, including the use of design goals. Helmut received a PhD in information science (Freie Universität Berlin, Germany) and master's and bachelor's degrees in computer science (Karlsruhe Institute for Technology, Germany).

Use Design Goals to Make Design Decisions Explainable and Defendable, page 57

Hillary Carey

Hillary Carey is the owner of Winnow Research and a PhD student in Carnegie Mellon's Transition Design program. Focused on understanding and finding patterns in what matters most to people, she uses creative, qualitative research methods and generative synthesis to define meaningful social change opportunities. Her dissertation inquiry examines how design approaches can contribute to racial justice. She is particularly interested in whether visions of better futures can engage and sustain people in the difficult transitions necessary to make change.

See Beyond the "Average" User, page 44

Holly Schroeder

Holly Schroeder (@314uxholly) is a UX researcher, instructor, artist, and writer. Her interest in people and what makes them tick transformed into a passion when she swiped her mom's RCA cassette recorder to play "interview." She believes accessibility is the secret sauce for designing awesome products. Her curiosity endures, fueling her passion for research. Currently, she works on a research team at a financial services company. She enjoys hiking, globetrotting via online events, painting, photography, and DIY projects.

Holly lives in St. Louis, Missouri, with her husband and their crew of rescue pets. Holly is copresident of the St. Louis Experience Design group (STLX) and is an executive board member of a11y STL. She has a bachelor of arts in psychology from the University of Missouri in St. Louis and a master of arts in nonprofit management from Washington University in St. Louis, as well as certificates in UX and web development.

Advocate for Accessibility, page 48

Ingrid Cruz

Ingrid Cruz is a people-focused product and design leader who builds innovative user experience and research organizations by inspiring teams to pursue excellence and continuous learning. She heads up the global UX and research organization at Ivanti, where they build tools for IT professionals who strive to provide a better employee experience and secure infrastructure. Ingrid is passionate about career development and mentorship. She currently serves as codirector of Product Hive, a Utah-based organization for product managers, designers, and researchers with more than eight thousand members and several chapters across the United States.

Don't Ask Users to Predict the Future, page 214

Jacqueline Ouifak

Jacqueline Ouifak is an experienced practitioner in user-centered design methodologies, with a focus on Lean UX and Agile UCD frameworks, who has grown practice teams and effectively scaled UX where it was needed the most. Her recent career focus has been on enterprise apps and nurturing a team of talented designers who can shine. She is currently leading a product design practice at CVS Health, a Fortune 5 company.

Always Do a Test of Your Test, page 180

James McElroy

James McElroy (@jamesmcelroy) stumbled into UX after getting talked out of architecture by a disgruntled architect he met at a BBQ. He pivoted to mechanical engineering with dreams of becoming an automotive engineer, until he met a human factors engineer at Bose Corporation months before graduation. James instantly fell in love with a profession he didn't know existed and joined Bose to prototype the world's first handheld touchscreen stereo interface (essentially Sonos on an iPhone a decade before the iPhone existed). He went back to school, becoming one of the first students in the Bentley University HFID program (where he now teaches part-time), and has built and led a number of teams focused on UX strategy, design, and research across diverse industries over the past 25 years. James lives with his family on an old farm in New Hampshire, where the family rooster likes to interrupt Zoom meetings.

You Can't Always Help Who You Want, page 92

Jen McGinn

Jen McGinn is a design and user research executive who has spent her career in enterprise software. Jen has cocreated and published new methods for persona development, heuristic evaluation, and Agile user research. Jen has coled the group mentoring program for the Boston UXPA conference since its inception and developed courses in Customer Experience and UX Processes for Bentley University and Brandeis University. Jen has a B.S. in Information Systems, and an M.S. in Human Factors in Information Design.

Expand Your Network Through Community Involvement, page 15

Jennifer Aldrich

Jennifer Aldrich (@jma245) is a UX-obsessed product designer and writer whose work has appeared in publications such as Creative Bloq, Startup Grind, UX Magazine, A List Apart, Yahoo News, Modus, Inside Design, and more. She has spent the last 15 years working for two SaaS start-ups. The first, Schoolwires, paved the way for more accessible online education opportunities for the next generation of students worldwide. The second, InVision, changed the way people design and collaborate cross-functionally in each of the Fortune 100 companies, supporting their ability to create the world's most impactful products. Jennifer is living her best remote worker life residing in and renovating a 1920s cottage in the country. In addition to designing, she loves spending time with her daughter, taking an obscenely large number of photos, and chasing around her three wild Bichon Frise puppies.

Mind Your Error Messages, page 130

Jesse Nichols

Jesse Nichols is a NN/g Certified user experience designer with over a decade of experience. At work, he designs interactions between people and systems that are productive, enjoyable, engaging, and satisfying. Jesse is happily married and lives in Augusta, GA. In addition to producing UX videos for his YouTube channel, BlackVNeckUX, he spends his free time on an odd collection of eclectic hobbies including voice acting, martial arts, rock climbing, chess, b-boying, sleight-of-hand magic, and knife throwing.

Turn Poorly Constructed Criticism into Actionable Feedback, page 72

Joe Sokohl

For more than 20 years, **Joe Sokohl** (@MotoUX) has crafted meaningful user experiences by using leadership, content strategy, information architecture, interaction design, and user research. He focuses on helping organizations understand who uses their products, how they interact with their products, and what benefits the company might realize by offering excellent experiences. He practices the mantra "UX for the rest of us." Lately he concentrates on productivity tools, web apps, and other internal interfaces that folks use in their day-to-day stuff. As Director of UX with Coforma, Joe offers change throughout diverse engagements, from healthcare provider intranets to government agency portals to financial advisor platforms. Focusing efforts on collaborative, distributed design, he provides strategic counsel for clients while keeping an ethical eye on user success. When not UXing, Sokohl passionately rides his motorcycle, writes about riding, history, and whiskey, and hikes with his wife, Karen, and their dog, Woody.

Don't Forget About Information Architecture, page 40

Joe Wilson

Joe Wilson's quest in marketing began when all good things began...in the '90s. Over the past two decades, Wilson has developed marketing strategies and digital solutions for companies like Gevalia Coffee, KIDZ Bop, Comcast, Drexel University, Accor Hotels, DuPont, Chemours, Axalta Coatings, Hush Puppies, Kindred Skincare, *winebasket.com*, Gypsy Warrior Clothing, and many other small to medium-sized businesses in the mid-Atlantic and NYC metro areas. In his younger years, you could find him touring the country in indie-punk bands and later as an acclaimed singer-songwriter. When he's not leading digital transformation as vice president of digital innovation for a large franchising organization in the home services industry, you can either find him writing poems about his daughter or writing songs...about his daughter.

Learn the Difference Between UX and UI from a Bicycle, page 76

John Yesko

John Yesko is a veteran experience design, creative, and digital business leader, now head of user experience and service design at Walgreens. He is responsible for the digital experience on Walgreens websites, consumer mobile apps, and in-store retail interfaces, as well as team member-facing digital. John also leads the omni-channel Service Design practice, with the mission of improving customer and employee experience across the 9,000+ Walgreens stores.

If You Show Something Shiny, They'll Assume It's Done, page 90

Jon Robinson

Jon Robinson (@jfarrellstudio) is a multidisciplinary creative director, UX designer, consultant, and professor of visual design. Based in St. Louis as a principal, experience strategy and design with Slalom—a global consulting firm that provides the world's top brands with solutions for business advisory, customer experience, technology, and analytics—Jon is also a member of the faculty at Lindenwood University's School of Arts, Media, and Communications and at the Washington University in St. Louis Sam Fox School of Design and Visual Arts.

Embrace Your Ignorance, page 151

Julia Choi

As a human factors engineer experienced in medical device development, **Julia Choi** has found a passion for advocating for users and their real-world needs, and in empowering design teams to value people over things. With training in cancer research, she was inspired by patients who vulnerably shared their personal experiences and continues to learn with diverse, international users in healthcare. She supports multidisciplinary teams to develop usable user interfaces including hardware, software, and instructions. For robotic-assisted surgery and interventional pulmonology, she has applied tools including contextual inquiry, task analysis, service blueprinting, service ecosystem mapping, risk analysis, usability studies, and datastories using qualitative and quantitative analysis. She has found meaning in creating safe spaces for those who are less heard or understood yet are critical to patient care through the messy and complex; this includes those on care teams, those who reprocess medical devices, and servicers of capital medical equipment.

Get Past Fear with Users and Design Teams, page 153

Julia Cowing

Julia Cowing (NYC) is a user experience researcher. She has worked at Citibank, Bloomberg, Mailchimp, and Google helping teams understand their customers by seeing problems through the perspective of their customers. The Customer Question Board is an outcome of her interest in creating frameworks to help organizations effectively conduct UX research in a systematic manner.

Plan User Research with the Customer Question Board, page 193

Julie Meridian

Julie Meridian (@juliemeridian) is a practicing user experience designer and artist in San José, California. She built upon the "learn by doing" philosophy of her alma mater, Cal Poly SLO, through working with the best at Adobe and LinkedIn. After leading design for multiple releases of Adobe Photoshop and Illustrator, and for LinkedIn Groups, Recruiter, and growth, she formed her consultancy, Make It Legit, which also includes service, enterprise, and healthcare design. She's mostly used her computer science degree to build rapport with developers...and to reassure designers that they don't need to code to be successful in design. Her studio art minor (and an editorial cartooning gig) blossomed into an art practice alongside her design. Sometimes these overlap during opportunities to create personas, customer journeys, and spot illustrations. Otherwise, she seeks creative cross-training through fine art as a resident artist of Local Color, Kaleid Gallery, and Works/San José.

Put On Your InfoSec Hat to Improve Your Designs, page 119

Kaaren Hanson

Kaaren Hanson builds robust design teams and creates customer-centered collaborative environments. Kaaren is obsessed with emotion and how companies can exceed customers' expectations. Currently, Kaaren is the chief design officer at JPMorgan Chase. Previously, Kaaren led design teams at Facebook, Medallia, Wells Fargo, and Intuit. While at Intuit, she launched and led the Design for Delight and Innovation Catalyst programs, ultimately creating a design-driven culture that led to world-class experiences. Her team's successes have noted in various media such as *Bloomberg Businessweek*, *Harvard Business Review*, *Creative Confidence* (Kelly & Kelly), and *Scaling Excellence* (Sutton & Rao). Kaaren has earned two CEO leadership awards and holds a BA from Clark University and a PhD from Stanford University.

Observed Behavior Is the Gold Standard, page 182

Katelyn Thompson

Katelyn Thompson (@kmtom) is a curious, observant user researcher with more than 10 years of experience in the field, including writing screeners for world travelers, pen enthusiasts, and those paying Uncle Sam. Katelyn holds an MS in human factors in information design from Bentley University. Katelyn is an avid geocacher and is always on the hunt for the next adventure.

The Right Screener Sets Up Your Recruit and Research for Success, page 199

Kathi Kaiser

Kathi Kaiser (@kathikaiser) is cofounder and partner at Centralis, a Chicago-based UX consultancy. For more than 20 years, she's led a top-notch team in creating outstanding user experiences for Fortune 500 companies, start-ups, and cultural institutions. When she's not at the whiteboard or in the lab, Kathi may be found observing users on boats, in museums, at train stations, and anywhere else that the digital and physical worlds collide. Kathi is a frequent speaker on topics in UX and has served as an adjunct faculty member at the Illinois Institute of Technology's Institute of Design. Kathi received her AB in psychology from Georgetown University and her MA in social science from the University of Chicago.

Know These Warning Signs of Information Architecture Problems, page 147

Kevin Lynn Brown

Kevin Lynn Brown is vice president of user experience and design, leading global UX teams and being responsible for product design across the entire corporation. His background is in fine art and graphic design. He has been designing websites and enterprise web applications for over 20 years.

Use Visual Design to Create an Eye Track, page 65

Kristian Delacruz

Kristian Delacruz is a designer who has worked for Discovery, Inc., led government digital transformation efforts at Booz Allen, and designed end-to-end user experiences at Asurion. He drove successful design solutions by utilizing user research, human-centered design, and product design to solve for service and digital challenges. For the past 10 years he's mentored and educated career changers and students at North Hagerstown High School, Shepherd University, General Assembly, and CareerFoundry. Kristian is continuously on the hunt to find the best food gems around the world and currently lives in Nashville, Tennessee, with his partner, Tyler, and their fur and furless children—Nash, Rufus, and Ember (the sphynx). His website is *delacruzdynamics.com.*

Design Mentorship Is a Lifelong Commitment, page 19

Kristina Hoeppner

Kristina Hoeppner (@anitsirk) is the project lead for the open source ePortfolio platform Mahara, working for Catalyst IT (*https://oreil.ly/yQyjV*) in Te Whanganui-a-Tara (Wellington) in Aotearoa, New Zealand. She traded hemispheres and careers in 2010 and enjoys supporting and working with the worldwide community of educators, learning designers and innovators, and portfolio enthusiasts. Kristina's responsibilities are manifold, among them working with the Catalyst UX team on realizing optimal implementation of new features in Mahara. Over the years, she has seen firsthand how UX helps shape products and make them better, learning constantly how to incorporate UX techniques into her client and community work.

Visualize Requirements During a Workshop, page 116

Kyle Soucy

Kyle Soucy (@kylesoucy) is the founding principal of Usable Interface (*https://oreil.ly/CYdu4*), an independent UX research consulting company specializing in user research and usability testing. Her industry-diverse client list includes companies such as Comcast, Intuit Inc., McGraw-Hill, Pfizer, and Wayfair, to name a few. She has spent over 20 years helping to create intuitive interfaces for a variety of different products, ranging from desktop and mobile apps to websites, medical devices, home entertainment systems, and even kitchen appliances. Kyle is the founder of the New Hampshire Chapter of the User Experience Professionals' Association (NH UXPA) and served as its inaugural president. She has also served as chair for the Philadelphia chapter of ACM's Computer-Human Interaction SIG (PhillyCHI). Kyle is very passionate about the continued growth of the UX community.

Data Alone Does Not Create Empathy—Storytelling Is Key, page 155

Lara Tacito

Lara Tacito has been a UX professional for over 10 years. She helps companies provide high-quality, easy-to-use digital experiences. She's worked at customer-centric companies like HubSpot, L.L.Bean, and Fidelity Investments. Her main skills include information architecture, systems design, and product design. She's currently a director of UX at HubSpot, where she's growing the user experience team. Before that, she helped build and grow HubSpot's design system, Canvas. She also evolved HubSpot's product architecture from a set of tools to a platform. At L.L.Bean she improved the searching and buying experience of the ecommerce website. And at Fidelity, she made personal investing an integrated financial application. She has a BS in finance from Lehigh University's College of Business. She currently lives and works in Massachusetts.

Create a Lasting Design System, page 108

Liz Possee Corthell

Liz Possee Corthell (@l_corthell) is a strategist, service designer, and futurist based in the Boston area. She works for Mad*Pow, a purpose-driven, strategic design consultancy, as an experience strategist. There, she works with clients in the healthcare industry to envision what the future of healthcare, well-being, and employee experience might be. Through research, service design, workshop facilitation, and strategic foresight, she helps tell the story of what the future could be and how we might get there. She is passionate about making the world a better place through good design and strategic foresight. When she's not working, she can be found cruising around town on her roller skates when New England weather permits, or knitting and crafting when it doesn't. She lives in Ipswich, Massachusetts, with her husband and their little orange cat.

Thinking About the Future Is Important for Any Design Process, page 34

Mac Hasley

Mac Hasley leads content strategy at dscout. Just as the best user researchers are passionate about supporting their users, Mac is passionate about supporting UXRs. You can read more of her work on the *People Nerds* blog (*https://oreil.ly/hsSVc*), where she works tirelessly to curate the best reads in research—including interviews with industry thought leaders, comprehensive methodological deep dives, and fascinating original studies. Outside of work, you'll find her writing creatively, cooking excessively, and drawing poorly.

You Need Good Planning for a Diary Study, page 207

Marino Ivo Lopes Fernandes

Marino Ivo Lopes Fernandes, PhD, is a visiting assistant professor of English at Bridgewater State University and content strategist for a government agency. Marino's writing and research have focused on the interface of language learning, writing pedagogy, and language access.

Align Your Tone, Voice, and Audiences, page 128

Marli Mesibov

Marli Mesibov (@marsinthestars) is the VP of content strategy at the digital UX agency Mad*Pow. Her work spans strategy and experiences across industries, with a particular interest in healthcare, finance, and education. She is a frequent conference speaker and a former editor of the UX publication UX Booth, and she was voted one of MindTouch's Top 25 Content Strategist Influencers. Marli can also be found on Twitter, where she shares thoughts on UX Design, content strategy, and healthcare. You can learn more about her and her work at her website, Adventures in Content Strategy (*http://marli.us*).

Design for Content First, page 126

Martha Valenta

Martha Valenta has been a UX consultant for the past 15 years. In that time, she has led UX on multimillion dollar projects for Fortune 100 companies. Prior to her career in UX, Martha was a graphic and web designer. In 2017 she cofounded the St. Louis Experience Design organization (STLX), which serves as a single source for experience design professionals in St. Louis. She has served on the boards of Service Design Network (SDN STL) and Gateway CHI and as chair of STLX 2019. She has presented at STLX, UXPA, and Design for America. She's also an artist. Some of her most inspired ideas for problem-solving happen when she's got a brush in her hand and a blob of wet paint on her clothes. @alwayscapturing is her Twitter handle and a nod to her insatiable curiosity and obsessive desire to untangle complex problems creatively through the lens of empathy.

On-Brand Whimsy Can Differentiate Your Mobile App, page 121

Martina Borkowsky

Since being confronted with a horrific custom software tool early in her career, **Martina Borkowsky** has been obsessed with Enterprise UX. She enjoys the complexity and high specialization of these tools, as well as the opportunity to help people feel happy and fulfilled at their jobs. Martina works as a UX design lead for McAfee, working remotely from the Netherlands, and is a Design Thinker by heart. The best thing about her job is meeting "her" users and learning about their demanding jobs, always trying to stay ahead of the bad folks out there to secure data and people. She does cybersecurity workshops for children, parents, and teachers as a member of the Online Safety for Kids team at McAfee. She loves speaking at schools and universities as a certified STEM ambassador in the hope of inspiring others for UX. She is a keen conference visitor, either as presenter or attendee, as she appreciates both roles equally.

Deliver Successful Products Through Common Success Metrics, page 165

Matt DiGirolamo

Matt DiGirolamo (@mdigirolUX) is a UX researcher, designer, teacher, and community leader in the Boston, Massachusetts, area. He is currently a senior user experience researcher at TextNow, where he is growing their research practice. He received his BS in Engineering Psychology and Biomedical Engineering from Tufts University and his MS in Information Architecture and Knowledge Management, User Experience Design from Kent State University. In 2018, Matt joined Lesley University as an adjunct faculty member teaching an introductory user experience course. Matt also volunteers on the UXPA Boston Board of Directors; he has served as the organization's president since 2019. In his spare time, Matt enjoys creating cocktails at his home bar, hosting trivia/game nights for friends, exploring great restaurants wherever he can, relaxing with his partner, Katherine, and playing tug-of-war with his Pembroke Welsh Corgi, Gilmore.

Scale Research Through Stakeholder Advocacy, page 170

Matthias Feit

Through his work as an application developer in a large network agency, **Matthias Feit** (@matthiasfeit) has had the (painful) experience that digital products and services developed without a deeper understanding of the user are rarely successful. He is all the more pleased today about the "Eureka!" moments with clients, the tectonic shift when priorities change from being features to customer needs. Matthias feels most comfortable in a role as a mediator in interface processes: between design and technology, customer and user, strategy and operational business. He advises clients with a focus on design systems, information architecture, and UX strategy and is co-organizer of Germany's Rhein-Main Area Meetup on Jobs to Be Done.

A Shared Vocabulary Can Increase Team Efficiency, page 132

Meena Kothandaraman

 With 30+ years of experience, **Meena Kothandaraman** has consulted to emphasize the strategic value and positioning of qualitative research in the creation of product, service, and space. Meena applies a credible, structured, and transparent approach to integrating human stories and anecdotes into mainstream processes. Meena is a founding member of twig+fish, a qualitative research and strategy practice based in Boston, Massachusetts, that espouses these research beliefs, while maintaining a utopic work-life balance. She also serves as lecturer in Bentley University's Human Factors and Information Design (HFID) graduate program. Over her 20-year tenure, her qualitative research curriculum has ably guided many leading research practitioners. She holds an MS in Information Resources Management from Syracuse University and a BCom in MIS from the University of Ottawa, Canada. Meena continuously draws inspiration from her passions: performing as a South-Indian Classical Violinist, pursuing culinary arts in a cooking show, and learning from her children!

Be Prepared When Practicing Ethnography, page 177

Megan Campos

 Megan Campos is the experience research director at Mad*Pow in Boston, Massachusetts, where she works across verticals including finance, healthcare, and education to deliver user insights and provide strategic recommendations for her clients. Stemming from her own interest in identity and her undergraduate years as a sociology major, Megan has become increasingly passionate about the importance of listening to diverse experiences when conducting research with users. Outside of work, Megan enjoys distance running, woodworking, knitting, and going on adventures with her dogs, Whiskey and Scotch. She is a dual citizen of Ireland and the United States. Megan has a BA in sociology from Dickinson College and an MS in Human Factors in Information Design from Bentley University. You can find her thoughts on UX and other matters on Twitter at @megangcampos.

Diverse Participant Recruiting Is Critical to Authentic User Research, page 143

Michael Hawley

Michael Hawley leverages a background in interaction design, usability, and design strategy to lead teams through the process of creating unique and engaging interactive experiences. He brings together experience in various disciplines, such as design research, content strategy, experience design, and visual design, in an environment that emphasizes creativity, teamwork, and advocacy for the person who will use the end solution. Michael is currently a senior experience consultant with ZeroDegrees and an adjunct professor at Bentley University. Formerly, he was chief design officer at Mad*Pow. Michael is an active participant in the research and design management community, contributing as a speaker, author, and mentor for UXPA, IxDA, Design Management Institute (DMI), and CHI. Michael holds an MS in Human Factors in Information Design from Bentley University Graduate School of Business and a BA from the University of Michigan.

Assess Usefulness and Desirability Early in Product Development, page 184

Michelle Morgan

Michelle Morgan is part rebel and part nerd on a lifelong journey to design better experiences for everyone. She has been a commercial and institutional architect, a design instructor, a coworking space founder and owner, a start-up advisor, and a UX designer. Her current work is focused on making complex business data and information actionable and elegant. She's also passionate about the intersection of space (digital and physical), information, and human cognition. Michelle has brought her interest in spatial morphology from architecture to digital design and looks to inform design work with a variety of models of perception and cognition, cultural lexicons, and personal experience to bring delight and understanding to users. Her other interests include travel, books, social change, feminism, math, spreadsheets, business models, cooking, gardening, real estate, and blowing away clients with a mix of creative and analytical thinking.

Make Learning a Part of Your Design Process, page 94

Monet Burse Moutinho

Monet Burse Moutinho is a cultural anthropologist and qualitative UX researcher. Monet attended Georgia State University for her BA in Cultural Anthropology and began her Masters in Social Cultural Anthropology and Human-Computer Interaction at the University of Cape Town. As an academic, Monet completed ethnographic field studies in Brazil, Ghana, and South Africa. As an industry professional, she currently leads qualitative user experience research in the financial technology sector. Monet is passionate about diversity in the technology industry, international travel, and spending time with her family.

Build a Culturally Reflexive Professional Framework, page 145

Morgane Peng

Morgane Peng is responsible for the design vision and strategy of Societe Generale Corporate & Investment Banking. She delivers united and meaningful experiences with her team across Societe Generale products for start-ups, corporations, and financial institutions. Previously, Morgane adventured into various fields of consulting, financial markets, and tech. In her spare time, she's a gamer at heart and works on an indie video game successfully crowdfunded on Kickstarter.

Not All Interfaces Need to Be Simplified, page 87

Navin Iyengar

Navin Iyengar is a product designer at Netflix, where he has led design on experiences that entertain more than 200 million households worldwide. He joined Netflix in 2008 as a founding member of the product design team for Netflix streaming, and he has shipped many experiences across mobile, web, and television since then. Recently Navin was the design lead on the interactive film *Black Mirror: Bandersnatch*, which won Emmys for Best TV Movie and Creative Achievement in Interactive Media. Navin earned a BS in computer science from the University of Illinois at Urbana-Champaign and then did graduate work in 3D animation and game design at the Academy of Art in San Francisco. He lives in San Francisco with his partner, Nicki, and their kids.

Question Your Intuition and Design to Extremes, page 112

Priyama Barua

Priyama Barua (*priyamabarua.com*) is the experience strategy director of the eastern region at Merge. She and her team are focused on creating seamless and innovative experiences using human-centered and participatory approaches. Priyama has been instrumental in implementing transformative end-to-end customer experience and new business models for Fortune 500 companies, especially in the health and finance domains. She is passionate about building more bridges between design and business and getting people aligned around a vision of change. In past lives she's been a professional fashion photographer, an entrepreneur, and a professor. Her master's in design management and MBA in operations enable her to build solutions that are sustainable and scalable and are reflective of business priorities. Originally from the tea-growing region of Assam in India, she now lives in Boston, where you'll find her exploring nooks and crannies and sailing on the Charles.

Boost Your Emotional Intelligence to Move from Good to Great UX, page 2

Rachel Young

Rachel Young is a user experience researcher and designer with over 20 years of experience, spanning telecommunications, insurance, healthcare, and military applications. Rachel has conducted research and solved UX problems as both a consultant and a member of in-house practice teams. With an MS in Human Factors and Applied Experimental Psychology from the University of Illinois, she is fascinated by how people think, the things we inherently do and don't do well, and how to apply all of that to the design of complex systems. She is employed as a senior advisor of UX Product Design at CVS Health. Rachel lives in northwest Chicagoland, along the banks of the Fox River, with her husband and three children. She enjoys creating art and getting her hands dirty in the garden and can often be found enjoying the outdoors with her family.

Educate Your Product Team for Successful User Research, page 160

Reena Ganga

Reena Ganga (@reenaganga) is a design lead at IBM's Silicon Valley Lab, creating experiences for emerging technologies. Having led design initiatives for augmented reality, virtual reality, and artificial intelligence, she enjoys planning for the design problems of tomorrow as well as solving the challenges facing customers today. Reena embraces the ambiguity of designing for emerging tech and enjoys sharing her learnings as a keynote speaker at global events. As a former TV news reporter and anchor, Reena has made a career out of answering the question *Why?* It's this same innate curiosity that drives her passion for good user experience. She notices things, she's bothered by things, and she's compelled to make them better. Reena holds a BA in journalism from the University of Technology Sydney and a MA in international relations from the University of Chicago. Born and raised in Australia, she currently resides in sunny California.

Master the Art of Storytelling, page 11

Rich Buttiglieri

Rich Buttiglieri is a proven leader in user experience research and design with expertise at building, managing, mentoring, and directing teams to deliver products with world-class user experiences. He has a breadth of experience in product domains as well as product types (enterprise SaaS and consumer desktop, mobile and web applications, IoT, AR, consumer electronics, medical devices, and commercial/industrial electronics). He has built several UX practices from the ground up over the past decade and has led teams and organizations as they mature in their UX practices.

Know When and How to Build a Usability Lab, page 173

Scot Briscoe

Scot Briscoe (@scotbriscoe) is a digital creative director and UX practice leader, responsible for building the community of practice vision and direction for product design, digital creative, and digital execution teams. Scot has led efforts to transform traditional or heritage companies into digital-first, brand-forward companies by leading digital experience strategy, implementing digital best practices and design systems, and balancing lean organizational structure and scalability with measurable and data-driven creative executions. He has led significant change management within multiple design communities to unlock creativity and design products that make the lives of users more productive, easy, and inspired by establishing customer-centered design processes and a culture of empathy, collaboration, creativity, and execution.

Learn to Think like a Missionary, Not a Mercenary, page 85

Shanae Chapman

Shanae Chapman (@nerdydivadesign) is a senior ux designer, founder of Nerdy Diva, and UX adjunct professor. She has an MPS in Informatics from Northeastern University, Certificate in Leadership from MIT, Certificate in Web Design from Webster University, and BA in Communication with a Certificate in African American Studies from Saint Louis University. Shanae is passionate about anti-racism education, human-centered design, and cloud technology. She has over a decade of experience designing products and services for enterprise companies such as Boeing, IBM, Red Hat, MathWorks, and Akamai. She has taught UX and web development at Lesley University, Northeastern University, Maryville University, and Fisher College. Shanae is the founder of Nerdy Diva, an organization that focuses on creating diverse and inclusive products, services, and workplaces through research, design, training, and coaching. Learn more at *nerdydiva.com*.

Create a Design Portfolio that Gets Results, page 22

Shanti Kanhai

Shanti Kanhai (@slakanhai) holds a bachelor's degree in cultural anthropology from the University of Amsterdam and a master's degree in public administration from Leiden University. She has designed mobile and web applications for enterprise as well as for smaller technology start-ups in emerging and disruptive industries based on computer vision. During her time at United Airlines, she has helped design data-heavy digital products to serve the needs of employees in aircraft maintenance and technical operations. Throughout her career she has been an advocate for ethnographic best practices in the corporate industry. Outside of work, Shanti likes to travel, hike, and try out new recipes.

Bring Themes to Exploratory Research, page 149

Shipra Kayan

Shipra Kayan (@skayan) is an entrepreneur and designer dedicated to transforming the way we work together as a global community. Based on her core belief that every human is inherently valuable and capable, Shipra's vision is to create a world in which two people of any cultural or geographic origin can come together to collaborate and learn from one another. As a design facilitator, Shipra guides cross-functional and distributed teams to achieve internal alignment on the core problems they are trying to solve, so that they build products and processes that truly serve. Shipra believes that including all voices in the creative process generates better solutions than any single person could achieve on their own. She therefore teaches every team member, no matter their assigned role, to be a designer: to listen to multiple perspectives, challenge the status quo, and build a shared vision of what could be.

Align Your Team Around Customer Needs via Design Workshops, page 81

Skyler Ray Taylor

Skyler Ray Taylor is a multidisciplinary user experience manager leading highly effective teams of designers and engineers. He's a hardworking leader and is relentless in his drive to find innovative solutions to challenging problems through design.

Be Wrong on Purpose, page 106

Sonia V. Weaver

Sonia V. Weaver is a career software protagonist in the thrilling story that brings useful and usable products to market. As a UX practitioner with 25+ years in technical writing, business analysis, interaction design, and user research, Sonia now focuses solely on leading her company's research program and practice to drive data and insights into the teams that shape, design, and build software. Her early work at IBM, EMC, and Autodesk—coupled with formal education in technical communications and human factors—motivated Sonia to give a voice to those struggling to use Enterprise and IT products. By connecting the dots between the operators' experiences and company revenue, she enjoys proving that creating products people like to use is a winning business strategy. Sonia is a former Texan now residing in

New England for the four seasons and beautiful scenery and spending time outdoors with her family as often as she can.

Create a Truly Visible UX Team, page 32

Stephen Denning

Stephen Denning (@steve_denning) is an experienced UX consultant with research as his center of gravity. He has a passion for trying to understand how the mind works, how people think and behave, and how that insight can be used to create useful, usable, engaging, and accessible experiences. He has worked in the technology space for over two decades, via studies in computer science, human factors, and information design, and has worked with clients as diverse as the BBC, Amazon, Vodafone, the Scottish Government, NHS, Salesforce, and McLaren Automotive. As director of UX at User Vision (based in Edinburgh, Scotland), Stephen helps clients understand their customers more deeply through robust research, design better products through user involvement, and work more effectively through a strategic approach to human-centered design. Outside of work, Stephen loves most sports that require a helmet.

Apply the Butterfly Approach to Interviews and Testing, page 212

Susan Mercer

After starting a career in geophysics, **Susan Mercer** discovered that she enjoyed visually communicating her data more than she did the science. She has over 25 years of expertise in user experience, working as a developer, a designer, a consultant, and now an in-house researcher. She has 10 years of experience in conducting user research using a diverse set of research methods across industries. Her current passions include scaling quality research within an organization and operationalizing digital tools such as research repositories to provide relevant research insights to stakeholders whenever needed. Susan has her MS in Human Factors in Information Design from Bentley University and is now an adjunct professor in that program and a senior manager of research and usability at Tripadvisor.

Tell the User's Story via Effective Research Reports, page 220

Taylor Kostal-Bergmann

Taylor Kostal-Bergmann is a UX researcher at Honeywell with expertise in service design. She is passionate about delivering user-centered experiences and loves to educate others about applying design thinking to various parts of their everyday lives. Taylor is inspired by the concept that designers have a responsibility to make the world a better place and strives to ignite that spark in others. She has enjoyed working in the field of experience design for the past eight years since graduating from the Savannah College of Art and Design with a BFA in service design. Taylor has worked in a variety of roles, from design to strategy, and from consulting to in-house teams at small agencies and large corporations. She has experience in a variety of industries such as aerospace, advertising, building and construction materials, finance, and food and beverage. Currently Taylor lives in north Atlanta with her husband and two children.

Your Worst Job May Be Your Best Learning Experience, page 4

Theo Johnson

Theo Johnson (@theorocker_UX) is a lead UX researcher for Sony Music Entertainment, where he is building out and leading the UX research practice in data and analytics. He has a passion for utilizing intuitive research tools and his expertise in UXR to create successful UX research practices. As a widely skilled researcher and designer, Theo has led multiple UX projects in ecommerce, iGaming, video game user experience, and AR/VR technology. He received both his bachelor of arts in cognitive psychology and master of arts in human-computer interaction from SUNY Oswego. During his academia, he traveled internationally to present and publish two IEEE virtual reality video game UX projects. He also has won awards at startup competitions, such as the Playcrafting + Microsoft Global Game Jam, for creating an artificial intelligence tabletop game. Theo also enjoys bowling, playing guitar, singing, and writing music in his free time.

Design Thinking Workshops Will Change Your Process, page 114

Thomas Yung

Thomas Yung is a UX designer, researcher, and developer with over six years' experience in the healthcare industry. He helps bring ideas to life using user research methods, an evidence-based process, an experimentation mindset, and a creator and maker skill set. He enjoys walks with his dog, drawing, and playing guitar. He is an advocate for diversity, inclusion, equality, and climate justice. He has lived in Asia, Africa, and Canada/the US and considers himself to be a citizen of the world.

You Don't Need a Lot of Money to Recruit Participants, page 204

Tim Heiler

Tim Heiler (@timheiler) helps groups make better decisions through structured conversations. His work is focused around cognition, behavior, and team psychology, helping enterprises leverage design to its full strategic potential. In addition to his current role leading design systems initiatives and software projects for large clients, Tim is involved in curating, organizing, and hosting several technology- and design-related conferences in and around Frankfurt am Main, Germany, including World Usability Day Frankfurt, Future Tools Now, and Themen der Zukunft. He currently holds the position of design director at the strategic design agency iconstorm and teaches design occasionally. Tim believes in the power of collective learning, and you will find him around the conference, meetup, and BarCamp scene in and around Europe. He loves a good bike ride and entertains a lifelong passion for electronic music in his free time.

Use Object Mapping to Create Clear and Consistent Interfaces, page 67

Todd Zazelenchuk

For the past 20 years, **Todd Zazelenchuk** has led UX design projects and teams in both industry and academic environments, including Intuit, Whirlpool Corporation, Plantronics, and Indiana University, where he earned his PhD in instructional technology and human-computer interaction. He has shipped successful hardware and software products, authored peer-reviewed articles, contributed book chapters, presented at international conferences, and received design and utility patents for his work across multiple industries. Currently, Todd is enjoying Austin, Texas, where he is the design director at Dealerware, an industry-leading software solution in the mobility space. Add Todd on LinkedIn (*https://oreil.ly/xKjLp*), and follow him on Twitter @ToddZazelenchuk.

Improve Usability Testing with Task Cards, page 210

Tripta Kumari

Tripta Kumari is a UX and interaction designer based in the UK. Her 20-year career has involved work across voice, mobile, web, and other platforms to champion the needs of the user, creating experiences that feel natural and enjoyable. She's worked with companies such as Nuance, Intel, Skype, Sony Ericsson, Vodafone, BT, John Lewis, Emirates, and ING. Her website and portfolio are at *triptakumari.com*.

Be Your Own Project Manager, page 100

William Ntim

William Ntim is a senior UX designer at PayPal. He has spent nearly a decade leading, creating, and launching design solutions at small- to large-sized corporations, including The Home Depot, HostGator, Ellie Mae, Farouk Systems, Zurvita, and BoomWriter. William is a critical thinker with a frontend development background, an author, a founder, a speaker, and a coach. In his downtime, he enjoys mentoring designers on best practices and the realities of being a UX practitioner.

Don't Perform a Competitive Analysis Before Ideating, page 123

Yingdi Qi

Yingdi Qi works on the product localization strategy team at Google. Her passion is the convergence between culture, design, technology, and international business. She enjoys using analytical and creative thinking to help products communicate internationally and enhance people's way of living through localization, international product design, and cultural adaptation.

Design Meaningful International UX, page 96

Index

A

A/B testing, 194

accessibility, 48-50, 52, 53

  (see also inclusive design)

Ackoff, Russell, 59-60

actionable feedback, 72-73, 166

actionable results from research, 155, 222

advice, asking for, 86

advocacy for value of UX team

  by enlisting colleagues from other
    teams, 17-18, 86, 170

  with long-term strategic goals, 85-86

  with techniques for increasing visibil-
    ity, 32-33, 170-172

affinity mapping, 37, 194

affordances in design, 45

Agile teams, collaboration with, 168-169

alignment, brain's perception of, 63

ambiguity in language, clarifying, 133, 197

analysis

  of competition, 123-124

  of completed projects, 94

  critiques as, 70-71

  and information architecture, 40-41

  of themes, 114, 150

  in UX design, 59, 70

anchoring bias, 28

animations, brain's perception of, 64

apps, special interactions for mobile,
  121-122

Arango, Jorge, 41

artboard approaches, limitations of, 43

assertiveness, 10, 47, 48, 164

assistive technologies (ATs), 51-52

assumptions in design work

  and international users, 96-97

  and people on the margins, 44-45

  questioning-based approach for chal-
    lenging, 151-152

  specifications document for avoiding,
    55-56

  wrong choices for challenging, 107

Astrophysics for People in a Hurry
  (Tyson), 151

ATs (assistive technologies), 51-52

attention spans, 26

attitudinal questions, 194

audience, target

  aligning tone and voice to, 128-129,
    131

  catching attention of, 11

  diversity of (see diverse populations,
    designing for)

  identifying content needs of, 126-127,
    131

understanding needs and intentions of, 138

and user experiences (see user experiences (UX))

user interfaces (UI) as basis for, 76-78

customer feedback surveys, 166, 194, 196-198

Customer Question Board in user research, 193-195

customer support as UX resource, 175-176

customer validation for design ideas, 80

D

daily work journal, keeping, 94

data

    customer support as, 175-176

    for identifying patterns of problems, 31

    and research (see research)

Dator, Jim, 34

decisions, defending with goals, 57-58, 79, 222

democratization of research, 172

demographics of participants, 143-144, 199, 202

design complexity/simplicity level, 87-89

design concepts

    avoiding competitive analysis with, 123-124

    creating multiple options for, 110-113

    creative risk-taking with, 102-103

    early-stage focus for, 184-186

    pitching (see pitching design concepts)

    and projects (see projects, design)

design decisions, user-centered, 123-124

design documentation specifications, clarity with, 55-56

design features as blinders, 30

design goals

    for communicating about problems and solutions, 79

    for defending decisions, 57-58

Design Justice (Costanza-Chock), 44

design portfolios, 16, 22-23, 122

design process, 82

    (see also design concepts)

    as advocacy, 86

    asking "Why?" in, 138-139

    communicating timeline of, 90-91

    with design goals, 57-58, 79

    emotional aspects of, 2-3, 164

    future scenarios and, 34-35

    learning as part of, 94-95

    object mapping in, 60, 68

    as open, 89

    pitching, 79-80

    reflexive and less-visible considerations for, 145-146, 164

    timeline of, 90-91, 100-101, 154, 167-169, 185

design systems, creating and maintaining, 108-109

design thinking workshops, 98, 114-115

design thinking, root of, 139

design workshops, planning and facilitating, 81-82, 114-117

developers, collaboration with, 32, 122, 166, 168

diary studies, 194, 207-209

digital platforms, portfolio templates on, 22

disabled people, designing for, 48-50, 54

    (see also inclusive design)

disaffordances in design, 45

discussion channels on UX design and research, 32

discussion, providing opportunities for, 27, 32, 71, 75, 82, 84, 150

diverse populations, designing for

    with diversity in research, 46, 143-144

    with diversity in UX teams, 46-47, 164

    as inclusion (see inclusive design)

    information architecture for, 41

with people on the margins, 44-45
  by recruiting diverse participants,
    143-144
diversity, defined, 46
documentation of design specifications,
  clarity with, 55-56
Domino's pizza, customer experience of,
  78
Don't Make Me Think (Krug), 110

E

education and training (see learning, culture of)
efficiency, increasing, 9, 108-109, 132-134,
  171
Einstellung effect with familiar solutions,
  110
email lists as networking tool, 16
emotional and mental needs of participants, 141, 153-154
emotional intelligence (EI), building, 2-3
emotions
  controlling, 2, 3, 73
  making connections with, 11
  in user personas, 157-159
empathy, building
  with customers, 138
  with disabled people, 49
  emotional intelligence for, 2
  with international users, 96-97
  with questioning mindset, 152
  with stakeholders, 154-156
  storytelling for, 154-156
  UX professionals and, 72
end-to-end customer experiences
  designing for, 30-31
  journey mapping of, 27-29, 31
engagement arcs of ethnographic observations, 177-179
enterprise products, complexity level for,
  87-89

error messages, crafting quality, 130-131
ethical considerations in research, 141, 146
ethnographic practices in user research,
  177-179
evidence-driven design approaches, 57-58,
  84
executive summary in research report, 221
experience mapping, 27-28
experimenting with design, 102-103
exploitation of design, protecting against,
  120
exploratory research, themes for, 149-150
eye tracking with visual design techniques,
  65-66, 194

F

fear, alleviating in participants and stakeholders, 153-154
feedback
  asking for, 3, 103, 111, 113
  constructive looping with, 144
  for different fidelity levels, 91
  diversity in, 47
  in implementing service design, 38
  for mentees, 20
  metrics for meaningful, 165-166, 171
  negative, benefiting from, 72-73
  as positive reinforcement for users, 62
  productive critiques as, 70-71
  questions eliciting actionable, 166
  for recruiters, 144, 203
filtering new content, 8
final moments in user experiences, importance of, 61-62
findability, 40-41
finishing projects, 10
framing device for focusing projects,
  104-105
fun and whimsy in design concepts,
  121-122
functional fidelity in prototypes, 42-43

future thinking and design process, 34-35

future-based questions, avoiding, 182, 214

G

Gamestorming (Gray, Brown, and Maca-nufo), 116

Gestalt laws (psychology principles), applying, 63-64

global audience, designing for, 96-97

glossary, common, 132-134

goals

for design, 57-58, 79

for flexible interviews, 212

long-term strategic, 85-86

in research reports, 220, 222

setting, 3

Gobet, Fernand, 110

Goleman, Daniel, 2

Gray, Dave, 116

groupings, brain's perception of, 63

H

hacking designs as purposeful practice, 119-120

headings in universal usability, 52

hireability, increasing, 16, 22-23, 122

homogeneity, limitations of, 143, 146

(see also diverse populations, designing for)

"How might we..." statements methodology, 115

human connections, making, 11, 26

human uniqueness, designing for, 53

hypothesis workshops, 81

hypothetical questions, avoiding, 214

I

IA (information architecture) (see information architecture (IA))

icon design, brain perception and, 64

ideas for concepts (see design concepts)

ideation sessions, 37, 58, 82, 150

ignorance as path to learning, 151-152

images versus text in universal usability, 51

in-situ logging for diary studies, 208

incentives for participants, 205

inclusive design

advocating for accessibility and, 48-50

basics of methodology of, 53-54

reflexive five-question guide for, 145-146

testing for, 120

universal usability and, 52

Inclusive Design Research Centre, 53

inclusive products, creating, 46-47, 53-54

information architecture (IA)

basics of, 40-41

signs of problems with, 147-148

information security (InfoSec) approach for anticipating problems, 119-120

information-seeking behaviors

problems with, 147-148

types of, 40-41

inputs and outputs of mobile devices, 122

interaction concepts, explaining and defending, 57-58

interactions, systems as web of, 60

interface expertise, 87-89

interfaces

complexity level of, 87-89

diversity issues with, 44-45

object mapping for clear and consistent, 67-68

usability studies for functionality of, 187-188

user (UI), 76-78

visual and functional fidelity of, 42-43

visual design of, 65-66

international audiences, designing for, 96-97

logo design, brain perception and, 64

M

Macanufo, James, 116

mapping
  affinity, 37, 194
  journey or experience, 27-29, 31, 37, 38, 81-82
  mind, 150, 158
  with objects, 60, 67-68

maturity models, 170

Mauborgne, Renée, 123

McLeod, Peter, 110

measurable outcomes in opportunity statements, 105

memory recall, timing and intensity in, 61-62

mentoring and mentors, 19-21, 86

method decision framework, 171

methodology section of research report, 221

metrics for meaningful feedback, 165-166, 171

microaggressions in UX design, 45

microexperiences, identifying, 29

Microsoft, inclusive design at, 54

mind mapping, 150, 158

mindfulness of moderators, 218

mistakes, space for making, 103

mobile devices, designing for, 120, 121-122

moderators in usability research
  answering questions with questions, 161, 219
  mental and emotional factors for, 141
  observing IA problems, 147-148
  probing for root causes, 187, 219
  quiet patience of, 187, 219
  using "Psychologist Voice", 218-219
  with varying degrees of control, 149-150

The Moderator's Survival Guide (Tedesco and Tranquada), 141

Molich, Rolf, 141

N

narrative scripts, 158

navigation structures
  effect of errors on, 131
  problems with, 147-148

negative feedback, benefiting from, 72-73

negative final impressions, avoiding, 61-62

networking through community involvement, 15-16

news aggregator apps, 8

newsletters for UX advocacy, 32, 33

nomenclature, problems with, 147-148

nonusers as participants in user research, 191-192

note-taking, 75, 94, 162, 178, 183, 213

O

object mapping
  for clear and consistent interfaces, 67-68
  for sensemaking with complex systems, 60

Objectives and Key Results (OKR) for setting goals, 3

objectives for solutions in critique process, 70-71

observation rooms of usability labs, 173-174

observations
  ethnographic open-ended, 177-179
  of user research sessions, 147-148, 162, 168-169, 173-174

observed behavior versus self-reporting of people, 182-183, 194

OKR (Objectives and Key Results) for setting goals, 3

one-way glass, pros and cons of, 174

product teams
  embedding research in, 171
  and user research, 160-162
product-market fit, 184
products
  assessing usefulness and desirability in,
    184-186
  assessments early in development of,
    184, 187
  enterprise, complexity level for, 87-89
  inclusive, 46-47, 53-54
  legacy, strategies for, 98-99
  localization of, 97
  metrics to measure success of, 165-166,
    171
  UX design in development of, 17
professional users, designing for, 87-89
profit margin, 14
profits, business, 13
project management, plans for, 100-101
projects, design
  anticipating weaknesses in, 119-120
  plan template for management of,
    100-101
  problems to solutions process for, 23
  reviewing and reflecting on, 94-95
  starting (see starting projects)
properties of objects, 67-68
prototypes
  functional as well as visual fidelity for,
    42-43
  localization with, 97
  for mobile inputs and outputs, 122
  testing of, 97, 180-181
  usability research with, 188
Proximity, Gestalt Law of, 63
"Psychologist Voice" for moderators in
  usability research, 218-219

Q
qualitative and quantitative questions, 194

qualitative and quantitative studies, sample sizes needed, 204
qualitative data from interviews, 150, 158,
  166
quantitative data from surveys, 33, 166,
  196
questioning mindset, 151-152
questions
  avoiding future-based hypothetical,
    182, 214
  for closed-ended responses, 196-198
  with Customer Question Board frame-
    work, 193-195
  for diary studies, 207
  with flexible interviews, 213
  guidelines for screener, 200
  "Is there anything else..." for unexpec-
    ted revelations, 216-217
  main research, defining, 215
  for open-ended responses, 196-198,
    199, 215
  on personalization, 185
  for surveys, 196-198
  "Tell me about..." for reliable answers,
    215
  "Why?," importance of, 138-139, 187,
    219

R
rapid user research methodology, 168-169
rational brain, dominance of, 2, 3
reader-based versus writer-based prose,
  128-129
readout sessions, 38
recommendations of research report, 222
recording user research sessions, 173, 213
recruiters, professional, 144, 188, 192,
  202-204
recruitment of participants
  agencies for, 144, 188, 192, 202-204
  from coworkers, 80, 189-190, 205

S

safety issues with participants, 140

sample size of participants, 204, 208

satisficing, concept of, 110

saying no, importance of, 5, 92-93

scalable text for universal usability, 52

schedule, project planning, 100-101

scheduling ethnography sessions, 177-179

screening for participant recruits, 143-144, 188, 192, 199-201, 202, 204-206

scripts for assessments, 180-181, 185

search terms, analysis of, 41

security concerns, 120

selectivity with new learning, 8

self-awareness, 2

self-management, 2

self-reporting, unreliability of, 182-183

selling design concepts (see pitching design concepts)

sensory experiences, 25-26

service blueprints, 37, 38

service design, implementation of, 36-38

shared flexible systems, 59-60

sharing new learning and information, 8, 172

signal-contingent protocol for diary studies, 209

"significant" design directions in design goals, 58

silos, avoiding, 83-84

Similarity, Gestalt Law of, 63

Simplicity, Gestalt Law of, 63

simulating customer experiences, 31, 43, 119

sketching by hand, benefits and approach, 74-75, 90-91

sketching with words, 135-136

sketching workshop, 81

skill set, increasing and diversifying, 5, 82

social awareness, 2

social gatherings, 15

social media, leveraging, 49, 176, 192, 202, 205

solution, not stopping at first, 110-113

solution-finding collaboration sessions, 169

solutions in critique process, objectives for, 70-71

speaking up about problems, 10, 47, 48, 164

specifications document, avoiding ambiguity in, 55-56

sponsorships within organizations, 18

spreadsheets in workshops, alternative to, 116-118

sprint planning, 94, 168, 189, 205

stakeholders

  alleviating fear in, 153-154

  building trust with, 80, 86

  communicating timeline of design process to, 90-91, 101, 154

  educating about UX process and value, 154, 171

  including as observers of usability sessions, 168-169

  legacy projects and, 98-99

  negative feedback from, 72-73

  pitching/selling design concepts to, 57-58, 79-80, 106-107, 113, 122

  using design goals with, 57-58, 79

  using research reports to present research to, 220-222

  using sketches to communicate with, 74-75, 90-91

  using stories to present research to, 155-156

  workshops with, 38, 98, 114-117, 195

standpoint theory in design, 44-45

starting projects, 104

  (see also design concepts)

  by accessing usefulness and desirability in products, 184-186

About the Editor

Dan Berlin

After providing technical support for hard-to-use systems for a number of years, **Dan Berlin** (@banderlin) discovered the world of user experience (UX) when he sat as a participant in a usability study. That's when he quit his job and went full-time to Bentley University to earn an MBA and an MS in Human Factors in Information Design. During his 13+ year UX career, Dan has lived the agency life and spent 10 of those years building the research team and practice at Mad*Pow, an experience design agency in New England. He recently started Watch City Research (*https://oreil.ly/EiLob*), a UX research consultancy outside of Boston, where he focuses on usability research in different product domains. He served on the UXPA Boston Board of Directors for eight years and served as the submission chairperson for their annual conference from 2012 to 2021. Dan also teaches in the Bentley University UX certificate program, is on the Board of Directors for the Waltham Land Trust, is advisor to the ΓΧ chapter of the ΣAM fraternity at Brandeis University, and can often be found kayaking along the Charles River or hiking with his dog, Shadow.

Leverage Your "Psychologist Voice" for Effective UX Research Moderation, page 218